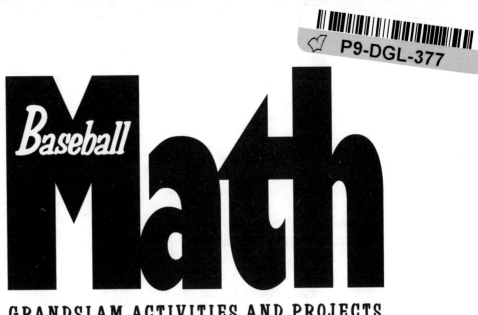

Baseball Math

GRANDSLAM ACTIVITIES AND PROJECTS
FOR GRADES 4-8

Christopher Jennison

GoodYearBooks
An Imprint of ScottForesman
A Division of HarperCollins*Publishers*

P9-DGL-377

Dedication

To Emily and Keith, my first and best teachers.

GoodYearBooks are available for most basic curriculum subjects plus many enrichment areas. For more GoodYearBooks, contact your local bookseller or educational dealer. For a complete catalog with information about other GoodYearBooks, please write:

GoodYearBooks
ScottForesman
1900 East Lake Avenue
Glenview, IL 60025

Design by Patricia Lenihan-Barbee.
Copyright © 1995 Christopher Jennison.
All Rights Reserved.
Printed in the United States of America.

ISBN 0-673-36122-5

6 7 8 9–PT– 02 01 00 99 98 97

Introduction

Welcome to **Baseballmath**, an activity and project book that taps one of a student's most involving interests. Assignments range from one-page skill reviews to projects designed to involve groups of students for several weeks. Throughout, the emphasis is on how and why, the essence of problem solving, as well as on communication and reasoning. Many activities include Challenge Problems, and numerous open-ended questions enhance critical thinking skills. Although the intended grade level range of the book is 4-8, there are no levels assigned to specific activities. Abilities and aptitudes can vary widely from grade to grade. The book's Table of Contents includes a listing of the skills employed in each activity and project.

Among the findings of an ongoing study supported by the Alfred P. Sloan Foundation was that sixth-graders prefer involvements that combine work and play. The students perceived such involvements as those that require the use of disciplined skills, such as sports, art, hobbies, and the learning of subjects like science and mathematics. John Dewey knew this when he said, ". . . experience is most rewarding when it involves the seemingly contradictory traits of rigor and playfulness."

The 1989 NCTM curriculum and evaluation standards provide a framework for developing curriculums that encourage

contextualized problem-solving and mathematical discourse. NCTM's 1991 standards for teaching mathematics emphasize solving nonroutine problems in meaningful contexts. A new middle school curriculum, Mathematics in Context, developed jointly by the National Center for Research in Mathematical Sciences Education at the University of Wisconsin-Madison and the Freudenthal Institute at the University of Utrecht, supports the vision of the NCTM Standards. The curriculum will explore problem situations based on real-world contexts, and emphasize the interconnectedness of mathematical concepts. And the comprehensive University of Chicago School Mathematics Project asserts: "Applications let more students get involved in mathematics activity and provide rich opportunities to teach both pure and applied mathematics," and "Virtually all students are able to learn to apply mathematics."

The lore and numbers of baseball lend themselves admirably to contextualized, real-world assignments. In the Morrison Elementary School in Philadelphia, Paula Goldstein recently posed this problem to her students: "On June 10, 1944, Joe Nuxhall pitched two-thirds of an inning for the Cincinnati Reds. But he had to get permission first from his high school principal. Joe was only fifteen years old and was the youngest player  ever to appear in the major leagues. How many outs did he record that day?"

In the pages that follow your students will project the value of their baseball card collections, compile averages, compute statistics, design a ballpark, compete in a fantasy baseball league, and much more. Activities in the book's first section are brief, one-page assignments; the projects in the second section of the book are more extensive and are intended for small- and large-group participation. In this section assignments range from estimating players' future performances to a classroom presentation of "Casey at the Bat." Cooperative learning strategies are modeled in most of the projects.

Use the activities for review, reinforcement, and enrichment. Don't hesitate to assign these activities to girls. The next time you go to a ball game, or watch one on TV, notice how many girls are in attendance. Girls are now included in Little League play, and softball and baseball are played by girls at many levels of competition. One of baseball's many attributes is that it can be played and enjoyed by almost everyone.

Mathematics is a means of investigation, a way of solving problems, and a way of thinking. It is connected to everything. Baseball and its associated activities and supporting context provide a meaningful connection. As Dewey proposed, the traits of rigor and playfulness need not be contradictory.

Acknowledgments
Randy Souviney, an old friend and former colleague, read an early version of this book and made some extremely helpful suggestions. A new friend, Jack Coffland, read a complete draft and offered some ideas for revision that were of enormous value. And more old friends: Bobbie Dempsey, Jenny Bevington, and Tom Nieman, the people behind GoodYearBooks, encouraged and supported me and exemplified the sort of author-publisher relationship that all writers cherish.

References
Alexander, Charles C. Our Game: An American History. *New York: Holt, 1991.*

The Baseball Encyclopedia: The Complete and Official Record of Major League Baseball, *9th ed. New York: Houghton Mifflin, 1988.*

Reichler, Joseph. The Great All-Time Baseball Record Book. *Rev. ed. New York: Macmillan, 1993.*

Siwoff, Seymour, et al. The Nineteen Ninety-Three Elias Baseball Analyst. *New York: Simon & Schuster, 1993.*

Thorn, John, & Palmer, Pete. Total Baseball. *3rd ed. New York: HarperCollins, 1993.*

Contents

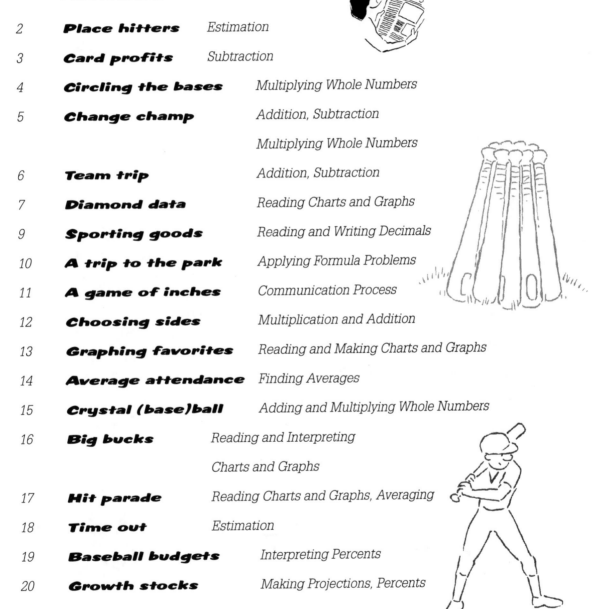

ACTIVITIES

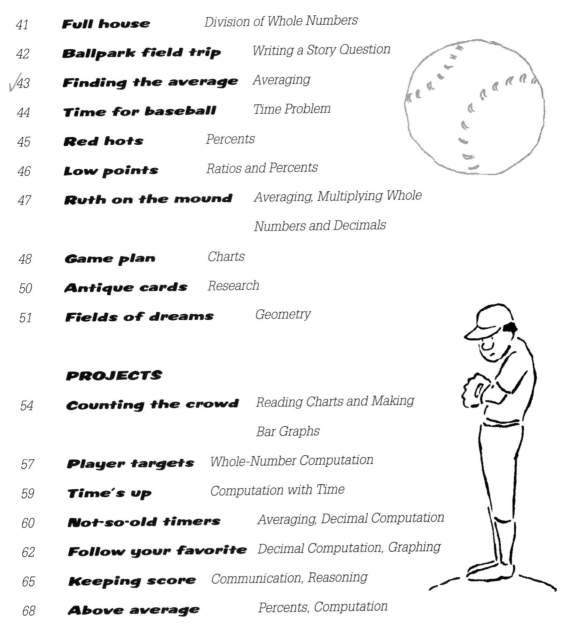

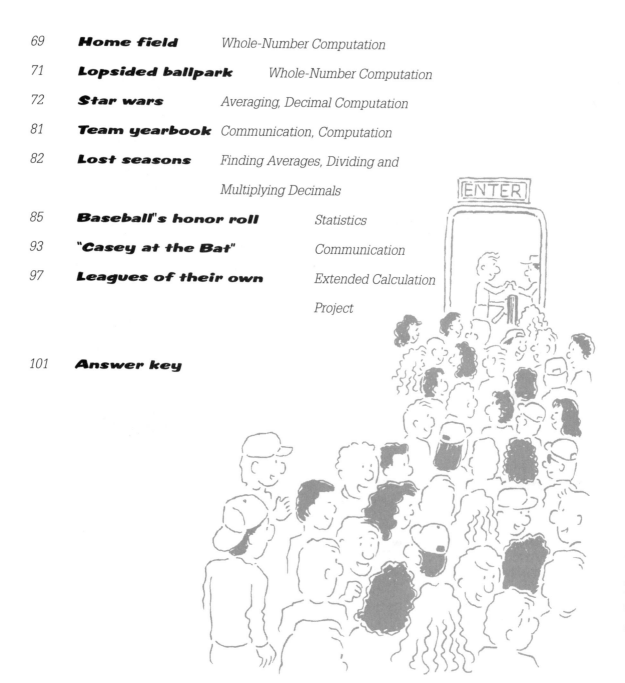

ACTIVITIES

Place hitters

The Valdez family went to a baseball game in June.

Circle the most sensible answers to the questions below.
Explain why the other choices are not sensible.

1. What was the cost of a ticket to the game?

$.80 $8.00 $80.00

2. How many hours did the game last?

3   30   300

3. How many baseball caps were sold by one vendor?

70   700   7000

4. How much did Mr. Valdez pay for a baseball cap?

$3.50 $35.00 $350.00

5. How many runs were scored in the game?

9   90   900

6. How many pitchers appeared in the game?

5   50   500

7. A trumpet player played the National Anthem before the game. How many minutes did the performance last?

3   30   300

8. How many players sat in the home team's dugout?

2   20   200

From *Baseballmath: Grandslam Activities and Projects for Grades 4-8* published by GoodYearBooks. Copyright © 1994 Christopher Jennison.

Card profits

Terry buys and sells baseball cards. She bought a Tony Gwynn card for 40¢ and then sold it for 60¢. A month later she bought the card back for 75¢ and then sold it a second time for 90¢.

1. When Terry bought the card for 40¢ and sold it for 60¢, did she make or lose money?

_____

How much?_____

2. When Terry bought the card back for 75¢ and sold it for 90¢, did she make or lose money?

_____

How much? _____

3. How much money did she make or lose in all?

_____

4. If Terry bought a card for 30¢, sold it for 50¢, and then bought it back for 70¢, how much would she have to sell it for to make a total of 50¢?

_____

Circling the bases

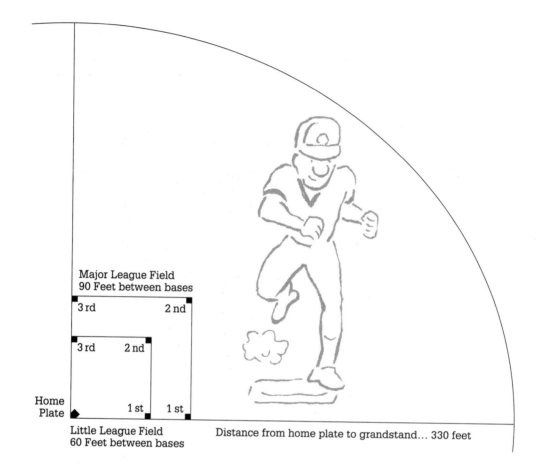

Major League Field
90 Feet between bases

3 rd   2 nd

3 rd   2 nd

Home
Plate   1 st   1 st

Little League Field
60 Feet between bases

Distance from home plate to grandstand... 330 feet

1. What is the total number of feet a player would travel around the bases if he hit a home run on a major-league field?

_____

2. What is the total number of feet a player would travel around the bases if she hit a home run on a Little League field?

_____

Challenge problem

3. If a player can run 10 miles per hour, how long would it take him to run from home plate to the grandstand along the first base line?

_____

From Baseballmath: Grandslam Activities and Projects for Grades 4-8 published by GoodYearBooks. Copyright © 1994 Christopher. Jennison.

Diamond data

Below and on the next page are data files showing what length baseball bat each player should use, according to the player's height and weight.

1. Greg is 5' 2" tall and weighs 109 pounds. What size bat should he use?

_____

2. Jose is 4' 7" tall and weighs 82 pounds. What size bat should he use?

_____

3. Based on your height and weight, what size bat should you use?

_____

4. In order to use a 32-inch bat, what is the maximum height a player should be?

_____

5. In order to use a 30-inch bat, what is the minimum weight a player should be?

_____

Baseball bat sizes

BOYS

	Batter's Height							
Batter's Weight in Pounds	3'5"– 3'8"	3'9"– 4'	4'1"– 4'4"	4'5"– 4'8"	4'9"– 5'	5'1"– 5'4"	5'5"– 5'8"	5'9"– 6'
under 60	27"	28"	29"	29"				
61-70	27"	28"	29"	29"	30"			
71-80	28"	28"	29"	30"	30"	31"		
81-90	28"	29"	29"	30"	30"	31"	32"	
91-100	28"	29"	30"	30"	31"	31"	32"	
101-110	29"	29"	30"	30"	31"	31"	32"	
111-120	29"	29"	30"	30"	31"	31"	32"	
121-130	29"	30"	30"	30"	31"	32"	32"	
131-140	29"	30"	30"	31"	31"	32"	33"	33"
141-150		30"	30"	31"	31"	32"	33"	34"
151-160		30"	31"	31"	32"	32"	33"	34"
over 160			31"	31"	32"	32"	33"	34"

6. If you are on a baseball team, check the heights and weights of some of your teammates. Are they using the correct length bat, according to the tables?

_____

Baseball bat sizes

GIRLS

Batter's Weight in Pounds	Batter's Height					
	3'10"–4'	4'1"–4'4"	4'5"–4'8"	4'9"–5'	5'1"–5'4"	5'5"–5'9"
under 40	26"	27"	28"			
40-45	27"	28"	29"	30"		
46-50	27"	28"	29"	30"		
51-60	27"	28"	29"	30"	31"	
61-70	28"	29"	30"	31"	32"	
71-80	28"	29"	30"	31"	32"	33"
81-90	29"	30"	31"	32"	33"	33"
91-100	29"	30"	31"	32"	33"	34"
101-110		30"	31"	32"	33"	34"
111-120		31"	32"	33"	34"	34"
121-130		31"	32"	33"	34"	34"
over 130			32"	33"	34"	34"

A game of inches

When he was in high school, the author of this book was a reporter for his town's newspaper, and he reported his high school's baseball games. He was paid 15¢ per column inch for his stories. The calculation was made simply by adding the length of the column or columns that the stories filled.

Times change, and reporters today are paid a lot more than fifteen cents an inch for their articles. Imagine that you are a reporter for your town's newspaper, and that you are paid $1.25 per column inch. Attend a high school baseball game and keep notes. Or, if possible, attend a professional game or watch a game on TV.

Write an article about the game. When you have finished, count the number of words in your article. Then look in your local paper and see how many words there are in one inch of a news column. Do this several times. Then find the average number of words in one inch of newspaper-column space. Figure out how many column inches your article would take up.

How much money would your article earn?   _____

Suppose you wanted to save $20 a week for a summer trip or to put in a college account. How many 200-word articles would you have to write each week to earn that much?   _____

If you wrote a 500-word article every day for the paper, how much money would you earn each week?   _____

Would you like to be a sportswriter when you get older? Why or why not?

_____

Choosing sides

Suppose you need names for three Little League teams. The players have suggested six names from which you can choose. One way of allowing more than one selection is by asking players to make first, second, and third choices. Ask seven players to select their first three choices for team names from the names suggested at the right. Next to each team name, write 1 each time a player selects that name as a first choice, 2 when it is the second choice, and 3 when it is the third choice. When you finish, you should have 21 numbers on your chart.

Teams	Student						
	1	2	3	4	5	6	7
Braves							
Athletics							
Pirates							
Angels							
Cubs							
Tigers							

Score points for the numbers above as follows:

> 1 = 5 points   2 = 3 points   3 = 1 point
>
> For example, a name that scored 1, 3, 2, 2, and 1 would be worth 17 points.

Write the total scores for each team:

Braves _____

Athletics _____

Pirates _____

Angels _____

Cubs _____

Tigers _____

Which are the three favorite teams?

_____

_____

_____

Why? _____

_____

_____

_____

From *Baseballmath: Grandslam Activities and Projects for Grades 4-8* published by GoodYearBooks. Copyright © 1994 Christopher Jennison.

Events

4	Player steals eight bases	15	for no money
7	No-hitter pitched	4	on the moon
22	Doubleheader is played	3	by a chimpanzee
5	Players agree to play	11	in one inning

1. This event will happen in 1998. The product of its two numbers is 21, and the sum is 10. What is the event?

No-hitter pitched by a chimpanzee

2. This event will happen in 2005. The product of its two numbers is 44. The sum is 15. What is the event?

_____

3. This event will happen in 2021. The sum of its two numbers is 26. The product is 88. What is the event?

_____

4. This event will happen in 2315. The sum of its two numbers is 20. The product is 75. What is the event?

_____

15 Crystal (base)ball

Big bucks

The graph below shows how many sports cards were sold between June 1991 and December 1992.

After studying the graph, answer the following questions.

1. Approximately how many cards were sold by the end of 1991?

_____

2. Approximately how many cards were sold by the end of 1992?

_____

3. Did the market for sports cards increase or decrease between 1991 and 1992?

_____

4. What was the approximate amount of the increase or decrease?

_____

5. Do you think the market for sports cards will increase or decrease in the next five years?

_____

Why? _____

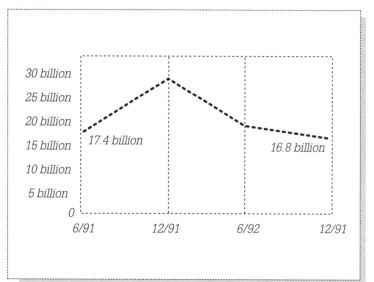

30 billion	
25 billion	
20 billion	
15 billion	17.4 billion 16.8 billion
10 billion	
5 billion	
0	

6/91   12/91   6/92   12/91

Challenge problem

According to Major-League Baseball properties, three of the companies that issue baseball cards—Upper Deck®, Topps®, and Donruss®—each control about 25% of the baseball card market. Fleer® has a 15% share and Score has 10%. In 1992, 16.8 billion baseball cards were sold. How many cards were sold by

Upper Deck® ? _____

Topps® ? _____ Donruss® ? _____

Fleer® ? _____ Score® ? _____

Hit parade

The Tigers, Cardinals, and Indians are keeping track to see which team makes the most hits in a ten-day period. This graph shows the total number of hits each team made.

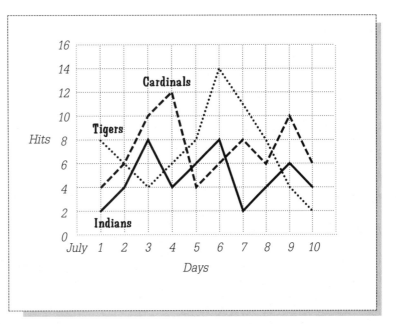

1. What was the first day each team was able to make six hits?

Tigers _____

Cardinals _____

Indians _____

2. On what day did the Tigers and Cardinals have the same number of hits?

_____

3. On what day were the most number of hits made?

_____

4. On what day were the fewest number of hits made?

_____

5. What was the average number of hits made by the Tigers during the ten-game period?

_____

6. Which team made the most hits during the ten-game period?

_____

7. Which team improved its hit total the most from one day to the next?

_____

8. Which team was most consistent throughout the ten-game period? (In other words, which team had the least variation between its high and low totals?)

_____

Time out

This activity will help you record how many minutes of commercials appear in each half hour of a televised baseball game.

1. First, estimate how many minutes of each half hour of a televised ball game are used for commercials. Record your estimate.

2. On the chart below, record the starting and stopping times of each commercial break during the game.

3. Make a bar graph to compare the number of minutes in the game used for commercials and the number of minutes in the entire game. Check to see how close your estimate was.

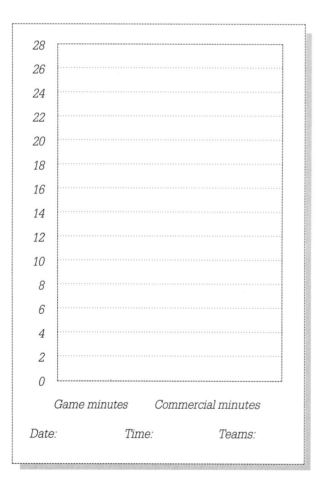

28
26
24
22
20
18
16
14
12
10
8
6
4
2
0

Game minutes   *Commercial minutes*

Date: *Time:*   *Teams:*

From *Baseballmath: Grandslam Activities and Projects for Grades 4-8* published by GoodYearBooks. Copyright © 1994 Christopher Jennison.

Baseball budgets

An imaginary baseball team budget is shown below. It is divided into percents.

Sometimes, instead of using a figure like 10% in a story, a newspaper reporter might say something like: "One of every ten team dollars is spent on maintenance." This is known as a ratio, a relationship, in this case, of 1 dollar to 10 dollars.
Keeping that in mind, read the following chart and answer the questions below.

1. What percent of the team's budget is spent on equipment?

_____

2. Write 6% as a ratio.

_____

3. One of every five budget dollars goes for

_____

4. Five out of every hundred budget dollars is spent on

_____

Expenditures by a baseball team

Player salaries	33%
Transportation	20%
Maintenance	10%
Insurance	8%
Equipment	7%
Concessions	8%
Administration	6%
Taxes	5%
Other	3%

5. Three dollars of every fifty is spent on

_____

6. Two out of every twenty-five budget dollars is spent on

_____

7. One of every three budget dollars is spent on

_____

8. Write a ratio in words to describe how budget dollars are spent on transportation.

_____

Growth stocks

If you collect baseball cards, chances are you're doing so because you think they'll be valuable in years to come. You could be right. Some cards issued in the last ten years have proven to be quite valuable (although few collectors are lucky enough to acquire cards such as the rookie cards of Nolan Ryan and Reggie Jackson). You can make a rough estimate of what some of your favorite cards and sets will be worth in years to come by using the forms and calculations below.

Card or set	
Year issued	
Cost new	
Estimated value now	
Years since issued	
Amount of increase	
Percent increase	
Average annual increase	
Value in one year	
Value in five years	
Value in ten years	

Begin with the name of the player or set. Then write down the year it was issued. You may not remember what the card or set cost you when you bought it, but you won't be too far off if you use 5¢ for a new card and $25.00 for a set. At the right is a sample calculation for Nolan Ryan's rookie card.

There are several magazines and price guides available that give present values of cards and provide information about the hobby in general.

Remember that the value of a card depends upon its condition, and that future values depend on how well a player continues to play and how well the overall market for baseball cards performs. Your estimates will be rough, but it's possible that there are some gems in your collection, so keep your cards in mint condition, and look forward to some pleasant surprises.

Card or set	Nolan Ryan
Year issued	1968
Cost new	.05
Estimated value now	$200
Years since issued	23
Amount of increase	$199.95
Percent increase	4000%
Average annual increase	174%
Value in one year	$286
Value in five years	$430
Value in ten years	$860

Inning time

1. Estimate the average playing time of one inning of a baseball game.

_____

2. Now watch or listen to a game and time each inning.

3. Make a bar graph below showing the time of each inning.

4. Do you think the first few or last few innings of a game are longer? Why do you think so?

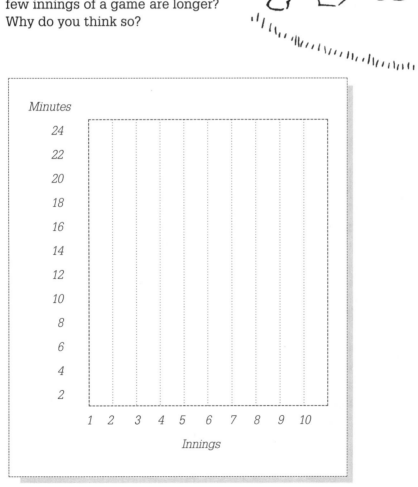

Minutes

24
22
20
18
16
14
12
10
8
6
4
2

1   2   3   4   5   6   7   8   9   10

Innings

Pitching pros

This activity helps you organize data, or information. The table on the right shows the names of several big league pitchers and the number of games they won in one season.

You can use a table to help you organize the data. First, tally, or write down, each pitcher's number of wins. Then count how many times an amount occurred. Write the frequency of each score in the table below. The mode is the total that occurred most often. Complete the table below. Tally each pitcher's total wins. Then count how often the amount occurred and write the amount under Frequency.

Name	Wins
Clemens	18
Cone	16
Gooden	18
Mussina	18
Johnson	15
Wilson	16
Rijo	19
Smoltz	14
Nagy	14
Abbott	18
Glavine	20
Perez	19

1. What is the highest number?

_____

2. How many pitchers won 16 games?

_____

3. Which pitcher had the most wins?

_____

4. Which is the lowest amount?

_____

5. Which amounts occurred the least often?

_____

6. What is the mode?

_____

Tallies	Frequency
14	Two
15	
16	
17	
18	
19	
20	

From Baseballmath: Grandslam Activities and Projects for Grades 4-8 published by GoodYearBooks. Copyright © 1994 Christopher Jennison.

Where they stand

Below are the standings in the Galactic League with just a week left in the season:

Team	Won	Lost	Pct.	Games Behind
Cosmos	87	69	.558	
Meteors	85	71	.545	2
Black Holes	82	75	.522	5 1/2
Lasers	81	76	.516	6 1/2

1. The season consists of 164 games. If the Cosmos lose all of the rest of their games, how many games must the Lasers win in order to tie for the lead, assuming the other two teams split their last games?

2. How many games must the Cosmos win to finish first, no matter what the other teams do?

3. The Cosmos have eight games left in the season. How many games do the Black Holes have left? If the Cosmos win just three of their last games, and the Black Holes win all of their remaining games, how will the two teams finish, assuming the other teams split their last games?

From Baseballmath: Grandslam Activities and Projects for Grades 4-8 published by GoodYearBooks. Copyright © 1994 Christopher Jennison.

Slugfest

Frances was reading an article about a home-run hitting contest. The article gave the following facts about how far each home run went.

Hank hit a homer 3 feet farther than Homer.
Champ hit a homer 3 feet farther than Slugger.
Aaron hit a homer 1 foot less than Hank.
Homer hit a homer 2 feet farther than Slugger.

When Frances turned the page, she was surprised to find that a page was missing. She decided to use the facts above to find out who hit the ball farthest. You can do the same using the table below.

Order of finish

First (Farthest)	
Second	
Third	
Fourth	
Fifth	

From *Baseballmath: Grandslam Activities and Projects for Grades 4-8* published by GoodYearBooks. Copyright © 1994 Christopher Jennison.

Tips for tips

You and your family have decided to spend the weekend in the city. You'll be going out to dinner, to a play or movie, and also to a ball game. It is customary to leave a tip for a waiter, a taxi driver, a bellhop, or anyone who provides a service. A standard tip is about 15% of the amount charged by a restaurant or taxi company. Before figuring a tip for each question below, round off the amount charged to the nearest dollar. For example, in question 1, you would round off the charge, $12.25, to $12.00. In question 3, you would round $7.80 to $8.00.

1. Cost of taxi from airport: $12.25
What should the driver's tip be?

_____

What is the total amount you should pay the driver?

_____

2. Cost of dinner: $63.38
What should the tip for the waiter be?

_____

What is the total bill for dinner?

_____

3. Cost of taxi from hotel to theater: $7.80
What should the driver's tip be?

_____

What is the total amount you should pay the driver?

_____

4. Cost of breakfast: $11.92
Cost of lunch: $27.61
What is the total amount you should pay for both meals?

_____

5. Cost of taxi from the hotel to the ballpark: $9.60
Cost of taxi from the ballpark to the hotel: $10.35
What is the total amount you should pay for both rides?

_____

Homer daze

Every week Tom and Luis keep track of how many home runs are hit in the major leagues during the previous seven days. The bar graph on the right shows the number of home runs hit during one week.

Use the bar graph to answer the questions.

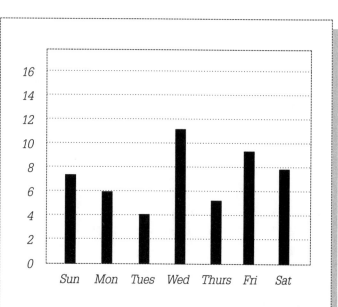

1. What multiples are used for the scale?

_____

2. On which day were the most home runs hit?

_____

3. On which day were the fewest home runs hit?

_____

4. On which day were 11 home runs hit?

_____

5. How many home runs were hit during the entire week?

_____

6. What multiples would you use on the scale if you wanted to show the number of base hits made on each day?

_____

Sunday attendance figures

7. Use the blank graph above to make a bar graph that shows the following information:

Sunday attendance figures

Braves 33,000	Yankees 19,000
Cubs 28,500	Expos 15,000
Giants 11,000	

From *Baseballmath: Grandslam Activities and Projects for Grades 4-8* published by GoodYearBooks. Copyright © 1994 Christopher Jennison.

Collecting cards

Franklin began a baseball card collection in January. At the end of March, he had 36 cards. At the end of July, he had 84 cards. He collects the same number of cards each month.

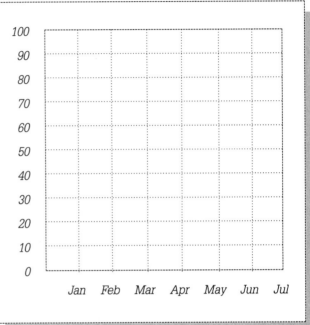

1. Make a line graph to show the growth of Franklin's baseball card collection.

2. How many cards did Franklin have at the end of each of these months?

January _____

February _____

March _____

April _____

May _____

June _____

3. At this rate, how many cards will Franklin have at the end of the year?

_____

4. If Franklin begins with 10 cards and increases his collection each month by 20%, how many cards will he have at the end of twelve months?

_____

Wait `til next year

Every team wants to improve its record from one year to the next.
When a team has a poor season, the players sometimes say,
"Wait 'til next year." Below are the 1991 and 1992 won-lost
records for the American League teams.

1. Which team had the most improved record from 1991 to 1992?

_____

2. Write that team's improvement as a percentage.

_____

3. Which team's record suffered the biggest drop from 1991 to 1992?

_____

4. What was that team's won-lost percentage in 1991?

_____

5. What was its won-lost difference in percent from 1991 to 1992?

_____

	1992		1991	
East	*W*	*L*	*W*	*L*
Toronto Blue Jays	96	66	91	71
Milwaukee Brewers	92	70	83	79
Baltimore Orioles	89	73	67	95
New York Yankees	76	86	71	91
Cleveland Indians	76	86	57	105
Detroit Tigers	75	87	84	78
Boston Red Sox	73	89	84	78
West				
Oakland A's	96	66	84	78
Minnesota Twins	90	72	95	67
Chicago White Sox	86	76	87	75
Texas Rangers	77	85	85	77
Kansas City Royals	72	90	82	80
California Angels	72	90	81	81
Seattle Mariners	64	98	83	79

Look up the records for these teams' 1993 and
1994 seasons. Which team made the steadiest
improvement over the four-year period? _____

Which team's record is declining most steadily? _____

From *Baseballmath: Grandslam Activities and Projects for Grades 4-8* published by GoodYearBooks. Copyright © 1994 Christopher Jennison.

Home-run showdown

1. Pretend that the day before the All-Star Game, the National League and the American League held a home-run hitting contest. Juan Gonzalez hit a 440-foot home run, and Fred McGriff hit a homer 380 feet. How much farther, on a percentage basis, did Gonzalez's homer go than McGriff's? (Round off your answer to the nearest percentage.)

23%   63%   16%

2. At that point in the season, Gonzalez's team, the Rangers, had played 54% of their games, and Gonzalez had hit 18 home runs. If Gonzalez continues to hit homers at the same pace, approximately how many home runs will he hit by the end of the season?

44   33   55

3. McGriff had 531 official at-bats during the 1992 season, and he hit 35 home runs. What was his home-runs-to-at-bats ratio?

❏ One homer for approximately 7 at-bats
❏ One homer for approximately 15 at-bats
❏ One homer for approximately 4 at-bats

4. In 1992 Gonzalez scored 77 runs, hit 43 home runs, and drove in 109 runs. How many runs did he produce in total? (Remember to subtract his homers from his runs and RBI totals so you don't count runs twice.)

188   207   143   129

5. Gonzalez played in 155 games in 1992. What was the ratio of his runs produced to games played?

❏ A little less than one run per game

❏ A little more than one run per game

From *Baseballmath: Grandslam Activities and Projects for Grades 4-8* published by GoodYearBooks. Copyright © 1994 Christopher Jennison.

Better than the boys?

For 11 years beginning in 1943, women played in the All-American Girls Baseball League. The league was made up of teams from around the midwestern part of the country, in cities like Rockford, Illinois; South Bend, Indiana; and Kenosha, Wisconsin. The teams played baseball, not softball, on regulation-sized baseball fields. During the first year, 4 teams played a 108-game schedule. How many games did each team play?

_____

Attendance during the first year was almost 108,000 fans. It was estimated that each team drew a larger percentage of their hometown's population than any major league team drew in its best season. In 1988 the New York Yankees' home attendance was 2,633,701 for the year. The population of New York City in 1988 was 7,350,000. What percentage of New York City's population attended Yankee games that year?

_____

Challenge problem

In 1943 the population of Rockford, Illinois, was about 85,000. How many Rockford fans would have had to attend their girls' baseball games that year in order to have attracted a higher percentage of the city's population than the Yankees did in 1988?

_____

From *Baseballmath: Grandslam Activities and Projects for Grades 4–8* published by GoodYearBooks. Copyright © 1994 Christopher Jennison.

The ballpark factor

Baseball teams are supposed to play better in their home ballparks. Here are the won-lost records, number of runs scored, and home runs hit by six teams in their home parks during the years shown in the first column.

Years	Team	Won	Lost	Runs	Homers	Opposition Homers
1934-92	Red Sox	2727	1898	24,204	4145	3687
1976-92	Yankees	789	554	6191	1294	1108
1968-92	A's	1136	860	8249	1677	1529
1966-92	Braves	1083	1059	9450	2046	2011
1916-92	Cubs	3231	2786	27,311	4456	4272
1962-92	Dodgers	1437	1045	9568	1605	1486

1. Which team had the best won-lost percentage in its home ballpark?

_____

2. Which team had the best home run average per season?

_____

3. Which team gave up the most home runs on an annual average basis?

_____

4. Which team's ballpark was the easiest to hit a home run in, based on an annual average of homers hit by the home team and its components?

_____

Challenge problems

5. How many times did the team with the best record at home win a league or division championship? (Look up this information in any baseball record book.)

_____

6. Do you think the teams with the best overall records have the best won-lost records at home?

_____

Why do you think so?

_____

_____

What else would you need to know in order to answer this question more completely?

_____

_____

From *Baseballmath: Grandslam Activities and Projects for Grades 4-8* published by GoodYearBooks. Copyright © 1994 Christopher Jennison.

Playing favorites

Which major league baseball player would you predict most people might pick as their favorite?

_____

Now it's time to find out if your prediction is correct. Select eight people to interview about their favorite players.

Name of Person	Age	Choice						

In the "Age" column, fill in one of the following words for each interviewee: Preteen, Teen, or Adult. At the top of each of the first three columns under "Choice," write the name of a player that you think will be mentioned most often. Leave four columns blank for now. Later, during your interviews, use these spaces to write in the names of players you had not previously written.

Conduct your interviews. Compare the predictions you wrote above with the data you collected. Who turned out to be the most popular player? Was any player mentioned by all three age groups? Did more people in your age group agree with your prediction?

_____

_____

_____

From *Baseballmath: Grandslam Activities and Projects for Grades 4-8* published by GoodYearBooks. Copyright © 1994 Christopher Jennison.

Bigger bucks

Starting in 1994, Darren Daulton, the Philadelphia Phillies catcher, will be paid $18.4 million for four seasons. What is his average salary per season?

_____

Assuming he plays an average of 130 games each season, what will his salary be per game played?

_____

At that rate, Daulton makes more in three games than Mickey Mantle made in his highest-paid season. Do you think Daulton is worth this kind of money?

_____

Why or why not?

_____

_____

_____

_____

Do you think any of the players earning millions of dollars a year are worth the money?

_____

Which ones?

_____

_____

Explain why.

_____

_____

_____

Most ballplayers' careers are over by the time they are 35 years old. But they could live for another fifty years after that, and not all of them are able to find jobs that pay well after they retire from baseball. Assume that a former ballplayer needs $50,000 a year to support himself and his family. Disregarding interest earnings and inflation, how much money will he need to support himself and his family at that rate for fifty years?

_____

If the player only played in the major leagues for 10 years, what would his average salary have to be in order to support himself and his family after he retired?

_____

From *Baseballmath: Grandslam Activities and Projects for Grades 4-8* published by GoodYearBooks. Copyright © 1994 Christopher Jennison.

Skill sharpeners

Six members of the Dolphins Little League team kept track of how many hours they practiced in one week. The chart below shows the number of hours each player practiced.

Student	Batting	Fielding	Bunting	Throwing	Pitching
Todd	5 1/2	3 1/6	2 1/3	0	1
Diana	1 1/6	1 1/6	1 1/2	3 2/3	0
Josh	2 1/2	1 2/3	1 3/4	3	0
Ruben	1 1/6	1	3/4	5 2/3	1
Benji	2 1/2	1 2/3	1	2 5/6	0
Connie	4 3/4	3 1/6	1 3/4	0	1

Choose an operation to solve each of the problems below. Then solve the problem.

1. How many hours did Todd spend batting and fielding?

_____

2. How much more time did Connie spend fielding than Josh?

_____

3. How much less time did Ruben spend bunting than Diana?

_____

4. If Benji spent the same amount of time fielding every week, how many total hours would he spend fielding after four weeks?

_____

5. How much more time did Ruben spend throwing than Benji?

_____

6. If Diana spent the same amount of time fielding and throwing each week, how many total hours would she spend on both after three weeks?

_____

Uniform numbers

Kelsey, Joe, Sonia, and Greg each have a favorite baseball uniform number. The numbers are 7, 11, 50, and 42. Use the clues below and the chart below to find each person's favorite number.

	Kelsey	Joe	Sonia	Greg
7				
11				
50				
42				

Clue 1
Joe's number is a multiple of 10. His number must be 50. Write YES next to 50 under Joe's name. Then write NO in the three other boxes under Joe's name. Write NO next to 50 under Kelsey, Sonia, and Greg.

Clue 2
Both Kelsey's and Sonia's favorite numbers are odd numbers. Only 7 and 11 are odd, so Greg's favorite must be 42.

Clue 3
Sonia's number is larger than Kelsey's.
What is Kelsey's favorite number?
What is Sonia's favorite number?

Complete the chart below to find each person's favorite number. Use the clues below.

Clue 1
Gary's number is odd.

Clue 2
Lionel's number is not a multiple of 5.

Clue 3
Chris's number is four times Ellen's number.

	Ellen	Chris	Gary	Lionel
4				
17				
25				
100				

From Baseballmath: Grandslam Activities and Projects for Grades 4-8 published by GoodYearBooks. Copyright © 1994 Christopher Jennison.

Long arms

Look at the table below. In the "Best" column, fill in the longest distance each student threw the ball.

Student	1st throw	2nd throw	3rd throw	Best
Carrie	25.73 m	26.35 m	26.4 m	
Roy	15.7 m	15.47 m	15.28 m	
Kyle	20.2 m	21.18 m	18.2 m	
Ingrid	15.07 m	26.34 m	26.24 m	
Eduardo	23.7 m	26.34 m	26.24 m	

1. Whose throws were the most consistent in length (who showed the least variation between throws)?

_____

2. Which student showed the most improvement from his or her first throw to his or her last throw?

_____

3. The school record for the baseball throwing contest is 39.68 m. How much farther would Carrie have to throw the ball to break the record?

_____

4. What is the difference between Roy's best throw and Eduardo's shortest throw?

_____

5. What is the total of Kyle's best throw and Roy's shortest throw?

_____

6. Whose throws are the least consistent (who showed the most variation between throws)?

_____

From *Baseballmath: Grandslam Activities and Projects for Grades 4-8* published by GoodYearBooks. Copyright © 1994 Christopher Jennison.

Picture players

You have been asked to put up 6 posters of your favorite players on a wall in the school library. Each poster is 2' 6" wide by 3' 3" high. The wall is 12' 6" high and 17' 4" wide.

1. What is the distance between the posters?

_____

2. If you want to hang 10 posters, all the same size, on the wall in 2 rows, how much space will be left between the posters in each row?

_____

3. How much space will you leave between each row and the top and bottom of the wall?

_____

From *Baseballmath: Grandslam Activities and Projects for Grades 4-8* published by GoodYearBooks. Copyright © 1994 Christopher Jennison.

Striking out

A good pitcher will strike out twice as many batters as he or she walks for a ratio of 2.00 to 1.00. The table below shows the strikeout and walk records of 6 pitchers.

Player	Strikeouts	Walks	Innings
Paul Blumenthal	46	32	109
Carlos Ortiz	29	18	67
Patty Boyd	31	19	88
Hector Jefferson	33	25	71
Phil Thompson	19	11	38
Karen Lind	27	24	52

1. Who has the best strikeout-to-walk ratio?

_____

What is that ratio, rounded off to two places?

_____

2. Who has the lowest strikeout-to-walk ratio?

_____

What is that ratio, rounded off to two places?

_____

3. Patty Boyd pitched a total of 88 innings. What was her average number of strikeouts per inning?

_____

4. Based on this average, how many strikeouts would she have after 134 innings?

_____

5. Paul Blumenthal pitched a total of 109 innings. What was his average number of walks per inning?

_____

6. Based on this average, how many walks would he have given up after 202 innings?

_____

7. Which pitcher had the best walks-per-inning ratio?

_____

What is it, rounded off to two places?

_____

8. Which pitcher would you like to have on your team?

_____

Why?

_____

From *Baseballmath: Grandslam Activities and Projects for Grades 4-8* published by GoodYearBooks. Copyright © 1994 Christopher Jennison.

Full house

1. In Lions Stadium there are 8 sections. Each is represented by a letter. If 1,537 people are sitting in section A and there are 53 rows in that section, how many people can sit in each row?

_____

2. If 3,060 people are sitting in Section H and there are 51 rows in that section, how many people can sit in each row?

_____

3. If there are 3,672 people sitting in Section C and that section has 12 rows, how many people can sit in each row?

_____

4. Section D has 24 rows. If a total of 2,520 people are sitting in that section, how many people can sit in each row?

_____

5. If there are 2,160 people sitting in Section G, and 72 people are sitting in each row, how many rows are in that section?

_____

6. If there are 840 people sitting in Section E and 42 people are sitting in each row, how many rows are in that section?

_____

7. A total of 5,488 people were seated in Section B on Sunday. That section normally holds 6,104 people in its 56 rows when it's full. How many people can sit in each row?

_____

8. A total of 4,740 people were seated in Section E on Sunday. That section holds 5,530 people in its 79 rows when it's full. How many people can usually sit in each row?

_____

Ballpark field trip

For a change, you will have a chance to write a question after you have been given the answer.

1. 564 students from Wilson School are going to the ball game on buses. Each bus holds 48 students. The answer is 12. What is the question?

_____

_____

_____

2. Now the answer is the remainder, 36. What is the question?

_____

_____

_____

3. The ballpark has 4500 seats. Each row holds 200 seats, except for the last row. Write a question so that the answer is the remainder, 100.

_____

_____

4. Write a question so that the answer is 22.

_____

_____

_____

5. Write a question so that the answer is 23.

_____

_____

_____

From *Baseballmath: Grandslam Activities and Projects for Grades 4-8* published by GoodYearBooks. Copyright © 1994 Christopher Jennison.

Finding the average

The Dynos baseball team played 20 games last season.

Five-Game Stretch	Scores	Average
First five games	3 11 9 4 3	
Second five games	2 19 3 7 9	
Third five games	6 12 1 5 1	
Fourth five games	4 13 4 9 5	

1. Find the team's average score for each five-game stretch. Use paper and pencil or a calculator.

2. Find the average number of hits for each player in the table below.

Player	Hits	Average
Ronnie	4 2 5 3 1	
Jenny	2 3 1 1 3	
Tom	5 1 1 4 4	
Bobbie	2 3 3 4 3	
Julio	4 4 5 2 5	

3. Find the average attendance for the five games in the table below.

Game 1	12,887
Game 2	11,040
Game 3	9,211
Game 4	4,530
Game 5	4,982

4. Find the average number of strikeouts for each player in the table below.

Player	Strikeouts	Average
Len	11 9 6 12 7	
Marian	7 9 9 10 5	
Sammy	10 5 8 15 7	
Chuck	12 4 7 11 6	
Stan	6 3 8 4 4	
Holly	8 5 9 10 3	
Michael	6 8 4 9 3	

Time for baseball

You will need a friend to complete this activity with you. Below is a map of the United States divided into four time zones.

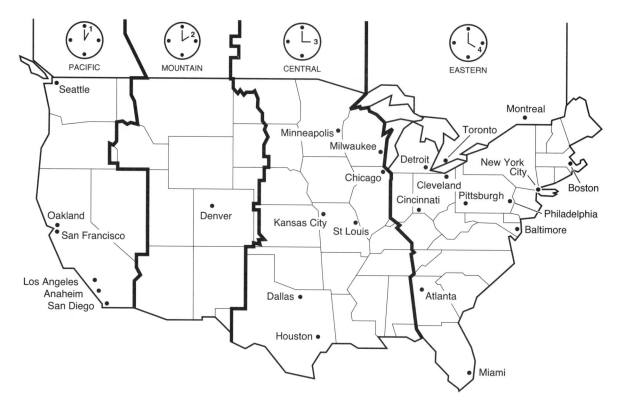

1. Write the names of the 26 cities on cards or small pieces of paper. Then, on 14 other cards or pieces of paper, write an hourly time, such as 3 p.m. Write a different hour on each card.

2. Mix each set of cards or sheets separately, and place them facedown in two stacks. Pick up a city card and a time card (for example, a card for Chicago and a card for 6 p.m.).

3. Ask a friend to pick up a city card. Based on the time it is in your city (6 p.m.), what time is it in your friend's city?

4. Reverse the process, with your friend selecting both a city and a time card and you choosing only a city card. What time is it in your city based on the time in your friend's city?

5. Keep track of how many correct answers each of you makes. This is an activity you can play many times, since the number of combinations of cities and times is high.

From *Baseballmath: Grandslam Activities and Projects for Grades 4-8* published by GoodYearBooks. Copyright © 1994 Christopher Jennison.

Red hots

Pretend that you are in charge of the concessions sold at your favorite team's ballpark. Concessions include everything from hot dogs to souvenir hats. You'll need to decide how much to charge for the items based on how much they cost you. For instance, suppose hot dogs cost you 20¢ each, and in order to cover your operating costs, you need to increase or "mark up" each hot dog by 200%. That means the cost to the customer must be 200% higher than what you paid for each hot dog. What will you have to charge?

_____

Now calculate the selling prices for the following items based on their costs to you and the percentage profit you must make in order to cover your costs of doing business.

	Cost	Markup	Price
Scorecard	.25	75%	
Hats	2.00	50%	
Autographed ball	3.50	40%	

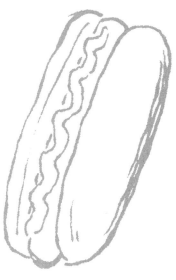

Challenge problem

For a little more figuring, provide the missing numbers for the items below.

	Cost	Markup	Price
Soda pop	.30		.90
Ice cream		50%	1.50

From *Baseballmath: Grandslam Activities and Projects for Grades 4-8* published by GoodYearBooks. Copyright © 1994 Christopher Jennison.

Low points

I'ts your sad duty to write a report about the worst hitters and pitchers in history. Their records appear below. To make the project a bit more interesting, some of the numbers are missing, and you'll need to fill them in, based on clues already given.

Name	At Bats	Hits	Ratio	Average	Ranking
Butter Fingers	1962	287	287/1962	.146	
O.U. Stink	2885	191	191/2885		
Shakee Batts	786		63/786		
Lowzee Hidder	6889		290/		
U.R. Badd		238	/4117		
Billy Goat	3998	328			
Hesa Whiffer			101/6632		
Bench Warmer	766	22			

	Won	Lost	Ratio	W–L Pct.	Ranking
Lefty Failure	19	188	19/188	.101	
Knuckle Flimsy	77	789			
Outa Control	32	411			
Bull Pen	66	972			
I.M. Crummy			22/619		
Uneeda Curve		452	71		

Now write a story that includes the ranking order of these players, and some information as to why they were able to last in baseball for so long. (You'll have to stretch your imagination for this part.) Then write about what one of the players did after he retired from the game. If you want, you could make up an interview with one of these players. The questions and answers could be quite creative!

From *Baseballmath: Grandslam Activities and Projects for Grades 4-8* published by GoodYearBooks. Copyright © 1994 Christopher Jennison.

Ruth on the mound

Everybody knows that Babe Ruth was a great slugger. And most people know that he was a good outfielder too. But not everyone knows that he was an outstanding pitcher at the beginning of his career, when he played for the Boston Red Sox. Below is the Babe's record as a pitcher. Using the years 1914-1917, try to calculate what his record would have been if he had been a pitcher for his whole career.

Ruth played a total of 22 years in the majors. Calculate an average of the pitching statistics Babe recorded between 1915–1918. Then you can estimate what he might have achieved over a 22-year career.

How many games would he have won? (Remember that this is just an estimate; there is no right answer.)

Where would this have put him on the all-time list of pitching victory leaders? (Look up the list in any baseball record book.)

What would his career earned run average have been?

How many strikeouts would he have had?

Where would this have put him on the all-time list of strikeout leaders?

What would have been his ratio of complete games to games started?

Year	Team	Won	Lost	Pct.	G	GS	CG	SO	SH	ERA
1914	Red Sox	2	1	.667	4	3	1	3	0	3.91
1915	Red Sox	18	8	.692	32	28	16	112	1	2.44
1916	Red Sox	23	12	.657	44	41	23	170	9	1.75
1917	Red Sox	24	13	.649	41	38	35	128	6	2.01
1918	Red Sox	13	7	.650	20	19	18	40	1	2.22
1919	Red Sox	9	5	.643	17	15	12	30	0	2.97

G Games   GS Games Started   CG Complete Games   SO Strikeouts
SH Shutouts   ERA Earned Run Average

Game plan

Below is the schedule for the New York Yankees in 1993. This is just an illustration to help you create a schedule for an imaginary league of four teams for a two-month period. Each team must play the other eighteen times; nine times at home and nine times on the road.

Create your own schedule using the forms on the next page. You should indicate home games by adding a color to the date square. Begin with one team and create its schedule with the other three teams. Depending on which months you use, each team will have six or seven days off, so space them through the schedule. You can also indicate day games and night games. Most teams play night games during the week and day games over the weekend, but you can arrange your schedule any way you want.

"1994 New York Yankees Baseball Schedule." Reprinted by permission of the New York Yankees.

From *Baseballmath: Grandslam Activities and Projects for Grades 4-8* published by GoodYearBooks. Copyright © 1994 Christopher Jennison.

Sun	Mon	Tues	Wed	Thurs	Fri	Sat

Month

Sun	Mon	Tues	Wed	Thurs	Fri	Sat

Month

Sun	Mon	Tues	Wed	Thurs	Fri	Sat

Month

Sun	Mon	Tues	Wed	Thurs	Fri	Sat

Month

Antique cards

One of this book's color inserts displays copies of baseball cards dating to the turn of the century. Collecting cards is hardly a recent phenomenon, but the hobby was not always pursued by young people. Some of the earliest cards were included in packs of cigarettes. In later years cards were put in gum packs and Cracker Jack boxes. Cards have also been distributed in vending machines, cereal boxes, loaves of bread, and boxes containing baseball equipment, and have even been given away by police officers.

The cards shown represent one gimmick to stimulate interest. Cartoons were popular, as were "triple folders," in which two player pictures framed an action photo. On the insert page are examples of a folder concept. The front of the card depicts a player in full length. About a third of the way down, the card was scored for easy folding. When the top one-third of the card is folded forward and down, another player is shown whose knees connect with the lower legs of the original player. Brief statistics are provided for each player on the back two-thirds of the card.

The insert is perforated and can be removed easily from the book. Perhaps you could research the career records of one or two of the players shown, or write a theme on how the old cards compare with modern ones. You might compare format and distribution approaches, as well as gimmicks.

From *Baseballmath: Grandslam Activities and Projects for Grades 4-8* published by GoodYearBooks. Copyright © 1994 Christopher Jennison.

Fields of dreams

The second color insert displays pictures of baseball stadiums that span almost exactly one hundred years, from the early 1890s to 1992. The ballpark at the top of the first page is the Polo Grounds, once located in upper Manhattan in New York City. The action shown took place during the 1950s, but the park was built in the 1890s and was remodeled and expanded several times during the early 1900s. It was torn down in 1964 to make room for apartment houses.

Beneath the Polo Grounds picture is a picture of Fenway Park, home of the Boston Red Sox. The park opened in 1912 and is still being used today. It too was expanded and remodeled over the years. The park's most distinctive feature is the left-field wall. It is called the Green Monster as a result of its paint job. Also, since it is so close to home plate, pitchers have grown to fear it.

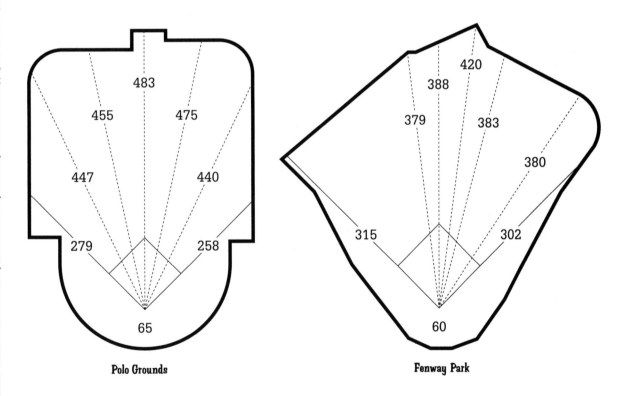

Polo Grounds   Fenway Park

The second page shows Cleveland Stadium, once called Municipal Stadium, and Oriole Park at Camden Yards, located in Baltimore. The Cleveland park was built in 1932 in hopes that the Olympic Games would be awarded to Cleveland, but the games went elsewhere that year. The Indians did not play there on a regular basis until 1947. It is a huge, and essentially featureless, park, and as such it anticipated the bland, circular stadia constructed during the 1960s and '70s. The Indians moved to a new ballpark in 1994, but the old stadium is still being used for football games.

Oriole Park at Camden Yards opened in 1992 and was designed purposely to replicate the look and feel of such classic ballparks as the Polo Grounds and Fenway Park. An old warehouse, seen on the right side of the picture, was not only saved but renovated, and its roof was used to support a bank of floodlights. The Orioles have enjoyed near-capacity crowds for every home game during the first two years of the park's existence, affirming, many believe, the value of a ballpark's "character."

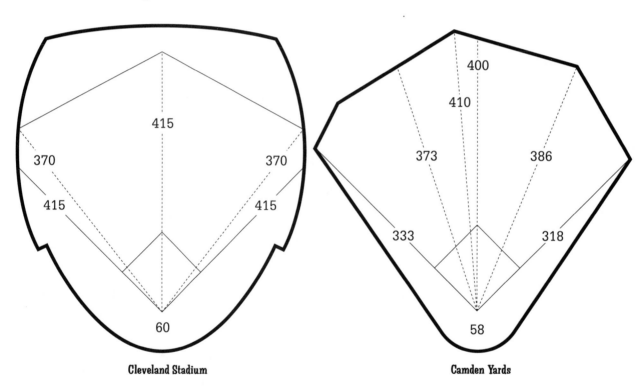

Cleveland Stadium

Camden Yards

You might undertake a research project that compared seasonal attendance figures for teams that moved from an old ballpark to a new one. What were the attendance figures for the Orioles during the years just prior to their move? Did the team play better in the new park? What effect might this have had on attendance figures?

From *Baseballmath: Grandslam Activities and Projects for Grades 4-8* published by GoodYearBooks. Copyright © 1994 Christopher Jennison.

PROJECTS

Counting the crowd

On the next page are the major-league attendance figures for the years 1990–1992 for each team. (The Marlins and Rockies are not included because they did not play their first seasons until 1993.)

Using the grid below, make a bar graph for three teams in each league.

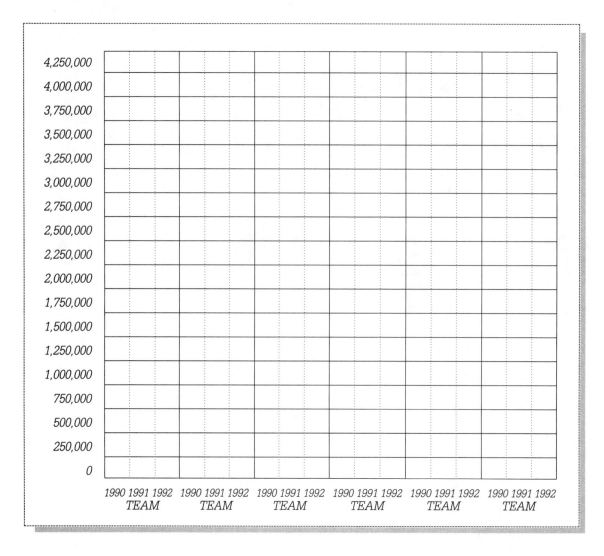

From *Baseballmath: Grandslam Activities and Projects for Grades 4-8* published by GoodYearBooks. Copyright © 1994 Christopher Jennison.

American League	1990	1991	1992	National League	1990	1991	1992
BAL	2,415,190	2,552,753	3,567,819	ATL	980,129	2,140,217	3,077,400
BOS	2,528,986	2,562,435	2,468,574	CHI	2,243,791	2,314,250	2,126,720
CAL	2,555,688	2,416,236	2,065,444	CIN	2,400,892	2,372,377	2,315,946
CHI	2,002,357	2,934,154	2,681,156	HOU	1,310,927	1,196,152	1,211,412
CLE	1,225,240	1,051,863	1,224,094	LA	3,002,396	3,348,170	2,473,266
DET	1,495,785	1,641,661	1,423,963	MON	1,373,087	934,742	1,669,127
KC	2,244,956	2,161,537	1,867,689	NY	2,732,745	2,284,484	1,779,534
MIL	1,752,900	1,478,729	1,857,351	PHI	1,992,484	2,050,012	1,927,448
MIN	1,751,584	2,293,842	2,482,428	PIT	2,049,908	2,065,302	1,829,395
NY	2,006,436	1,863,733	1,748,737	STL	2,573,225	2,448,699	2,418,483
OAK	2,900,217	2,713,493	2,494,160	SD	1,856,396	1,804,289	1,721,406
SEA	2,509,727	2,147,905	1,651,367	SF	1,975,528	1,737,478	1,560,998
TEX	2,057,911	2,297,720	2,198,231	TOT	24,491,508	24,696,172	24,111,135
TOR	3,885,284	4,001,527	4,028,318	ML	54,823,768	56,813,760	55,870,466
TOT	30,332,260	32,117,588	31,759,331				

1. Which league has the highest attendance figures?

_____

2. Which team showed the most consistent improvement over the three-year period?

_____

Can you think of one reason for this?

_____

3. Which team's attendance dropped most dramatically? Why do you think this happened?

_____

_____

_____

_____

_____

_____

4. Is the attendance trend up or down? Look at the graph and see if you can tell what the general tendency is from the heights of the bars and the directions they are indicating.

_____

_____

_____

_____

5. What factors do you think have the most influence on attendance figures?

_____

_____

_____

_____

_____

_____

_____

_____

6. Do you think more people will attend games in future years, or do you think fewer people will? Why do you think so?

_____

_____

_____

_____

_____

_____

_____

From _Baseballmath: Grandslam Activities and Projects for Grades 4-8_ published by GoodYearBooks. Copyright © 1994 Christopher Jennison.

Player targets

At the end of the 1993 season, Cal Ripken, Jr., of the Baltimore Orioles had played in 1,897 consecutive games. The record of 2,130 games was set by Lou Gehrig between 1925 and 1939. There are 162 games in a season. Approximately twenty games are played in April, thirty in May, thirty in June, twenty-five in July, thirty in August, and thirty in September. Estimate the month and year that Ripken will break Gehrig's record.

A goal for many hitters is 3,000 hits during their career. At the end of the 1993 season the following players were getting closer to their goal of 3000 hits:

Lenny Dykstra 1,110	Tony Gwynn 2,039
Eddie Murray 2,820	Ozzie Smith 2,265
Andre Dawson 2,630	Kirby Puckett 2,006
Paul Molitor 2,492	Wade Boggs 2,267
Ryne Sandberg 2,080	Don Mattingly 1,908

Which players have the best chance of reaching 3,000 hits?

_____

_____

_____

_____

Why do you think so?

_____

_____

_____

_____

List three players who are in the second or third years of their careers that you think have a chance to reach 3,000 hits. Calculate the number of hits these players have averaged per season so far, and then estimate how many seasons it will take them to reach 3,000 hits.

_____

_____

_____

A target for pitchers is 300 victories in a career. Twenty players have achieved this so far. Cy Young, who pitched for twenty-two years from 1890-1911, won a total of 511 games! Between 1891 and 1904, he averaged slightly more than 28 victories per season. Today's pitchers don't stand a chance of winning 500 games, but a few could win 300. List five below, making the same calculations and estimates you made for hitters. There are no right or wrong answers. Just have fun backing up your predictions.

_____

_____

_____

_____

From *Baseballmath: Grandslam Activities and Projects for Grades 4-8* published by GoodYearBooks. Copyright © 1994 Christopher Jennison.

Time's up

Some people think that baseball games are getting too long. There was a time, not so long ago, when games lasted between two and two and a half hours. Are games longer today? Below is a chart you can fill in over a twenty-game period. It will compare the lengths of games and if they were night or day games. The information you need is provided in your newspaper.

	Home game / Away game	Night game / Day game	Game Length		Home game / Away game	Night game / Day game	Game Length
1.				11.			
2.				12.			
3.				13.			
4.				14.			
5.				15.			
6.				16.			
7.				17.			
8.				18.			
9.				19.			
10.				20.			

What was the average length of the games?

_____

Were your team's home games longer or shorter than games they played away?

_____

Were night games longer or shorter?

_____

Do you think baseball games are too long?

_____

What could be done to shorten them?

_____

To make this project more interesting, you could create another chart and compare American League and National League games to see if there is any difference in the lengths of games. On this chart you would only list the game lengths. Was there a significant difference (fifteen minutes or more) on average, from one league to the other? If there was a difference, can you think of one reason for the difference?

Not-so-old timers

Lots of friendly arguments are caused when modern players are compared with so-called old timers. Below is a lineup of players who played during the 1950s and 1960s. Their best single season records are given, and their career records.

Player	Pos.	Best season						Career					
		AB	R	H	HR	RBI	AVG	AB	R	H	HR	RBI	AVG
Berra	C	584	88	179	22	125	.307	7555	1175	2150	358	1430	.285
Musial	1B	611	135	230	39	131	.376	10972	1949	3630	475	1951	.331
J. Robinson	2B	593	122	203	16	124	.342	4877	947	1518	137	734	.311
B. Robinson	3B	612	82	194	28	118	.317	10654	1232	2848	268	1357	.267
Banks	SS	617	119	193	47	129	.313	9421	1305	2583	512	1636	.274
Williams	LF	566	150	194	43	159	.343	7706	1798	2654	521	1839	.344
Mays	CF	580	123	185	51	127	.319	10881	2062	3283	660	1903	.302
Aaron	RF	629	116	223	39	123	.355	12364	2174	3771	755	2297	.305

Player		W	L	PCT.	G	IP	H	BB	SO	ERA	W	L	PCT	G	IP	H	BB	SO	ERA
Koufax	P	27	9	.750	41	323	241	77	317	1.73	165	87	.655	397	2324	1754	817	2396	2.76

On the form on the next page, write the names of the players you think are the best at their positions today. List their best single season records and then their estimated career totals. To make an estimate of a player's career totals, calculate their average statistics in each category per season, then multiply the average number by the number of seasons you think the player will remain in the majors. Remember that production usually drops in a player's last years, so you should probably use lower-than-average figures for a player's last two or three seasons.

How do your all-stars compare with the old timers?

_____

_____

_____

How many of your players were better?

_____

_____

_____

From Baseballmath: Grandslam Activities and Projects for Grades 4-8 published by GoodYearBooks. Copyright © 1994 Christopher Jennison.

		Best season							Career					
PLAYER	POS	AB	R	H	HR	RBI	AVG		AB	R	H	HR	RBI	AVG

	W	L	PCT.	G	IP	H	BB	SO	ERA	W	L	PCT.	G	IP	H	BB	SO	ERA

If you were to create a team from both lineups, how many old timers would be on it?

_____

_____

_____

Why?_____

_____

_____

Do you think modern players are better than the old timers?

_____

Why?_____

_____

_____

How many of baseball's most important records are held by players who played before 1970?

_____

_____

Challenge problem

Another measure of a pitcher's effectiveness that doesn't always appear in the record books is his number of strikeouts compared to the number of walks he gave up. During Sandy Koufax's best season, his strikeouts-to-walks ratio was better than four to one. What was his career strikeouts-to-walks ratio?

_____

Are there any modern players with ratios as good? What about Roger Clemens?

_____

Follow your favorite

Here's a chance to follow your favorite player over a long period. The chart on the next page provides enough space for sixteen weeks, but you can complete the project in less time, if you prefer. You'll keep a record of your favorite player's cumulative and weekly batting average, and then plot the information on a graph for easy reference (see page 00). The sample chart below will help get you started. To figure out your player's weekly average, you first find the difference between last week's and this week's hits. Then you find the difference between last week's at-bats and this week's at-bats. Divide the number of hits by times at bat to get the weekly average.

	Cumulative			Weekly		
	Hits	Times at bat	Avg.	Hits	Times at bat	Avg.
Week 1	7	21	.333	7	21	.333
Week 2	13	45	.289	13-7=6	45-21=24	.250
Week 3	20	66	.303	20-13=7	66-45=21	?

From *Baseballmath: Grandslam Activities and Projects for Grades 4-8* published by GoodYearBooks. Copyright © 1994 Christopher Jennison.

_____ **Stats**

Name of player

	Cumulative			Weekly		
Week #	Hits	At Bat	Avg.	Hits	At Bat	Avg.

 Follow your favorite

_____ **Batting average**

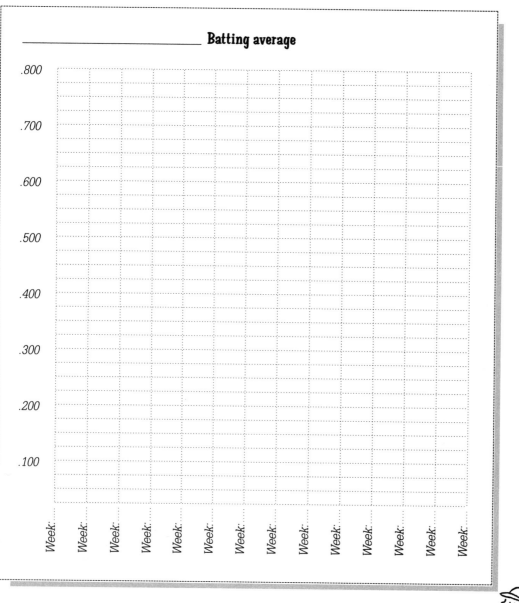

.800

.700

.600

.500

.400

.300

.200

.100

Week: Week: Week: Week: Week: Week: Week: Week: Week: Week: Week: Week: Week: Week: Week:

Follow your favorite

Keeping score

Keeping score makes a baseball game a lot more fun, and it's not as complicated as it looks. First of all, each player position has a number; not a uniform number, but a number used for scorekeeping purposes. These numbers are always the same: 1-Pitcher, 2-Catcher, 3-First Baseman, 4-Second Baseman, 5-Third Baseman, 6-Shortstop, 7-Left Fielder, 8-Center Fielder, 9-Right Fielder. Below are marks and symbols used for keeping score.

Single	—	Reached on error ... E		Stolen base ....SB		Balk ....BK	
Double	=	Fielder's choice ....FC		Sacrifice ......SAC		Strikeout ....K	
Triple	≡	Hit by Pitch ....HP		Sacrifice Fly ...SF		Walk ....BB	
Homerun	≡	Wild Pitch............WP		Passed Ball .....PB		Force Out ...FO	

Pretend that the scoring square or block is a baseball diamond, and that the lower-left corner is home plate, the lower-right corner is first base, the upper-right corner is second base, and the upper-left corner is third base. Below are four examples of scorekeeping:

E6   SB
(PB) —

In this example, the hitter got to first on a single, stole second base, got to third on an error by the shortstop and scored on a passed ball. Put a circle around a scoring play so runs on a scorecard will show up at a glance

FO
6-4

BB

In this example, the batter reached first on a walk, and then was out at second on a force play when the shortstop fielded a ground ball hit by the next batter and threw to the second baseman who made the putout.

5-3

In this example the batter hit a ground ball to the third baseman who threw to the first baseman for the putout.

SAC   =

SF8

On this example, the batter hit a double, reached third on a sacrifice by the next batter, then scored when the next batter hit a sacrifice fly to the center fielder.

Take a look at three complete
innings of scorekeeping. This will give you
a better idea of how an entire game would
look on a scorecard.

PLAYER	POS.	1	2	3	4	5	6	7	8	9	10	11
Wilson	2B	K		_E_								
O'Brien	LF	9		6-4 E5								
Chavez	3B	⊜		6-4-3								
Cole	1B	3-1		///								
Nieman	C	///	2									
Bevington	CF		6-3									
Dempsey	RF		2-_6_									
Lubin	SS		///	3								
Gamboli	P			⊖								
TOTALS RUNS / HITS		1 / 1	0 / 1	/								

From Baseballmath: Grandslam Activities and Projects for Grades 4–8 published by GoodYearBooks. Copyright © 1994 Christopher Jennison.

In the first inning, Wilson struck out, O'Brien hit a fly ball to the right fielder, and then Chavez hit a home run. Cole ended the inning with a groundball to the first baseman who threw to the pitcher covering first for the out. Be sure to make a mark in the box below so you don't start there in the second inning. Nieman started the second inning by popping out to the catcher, then Bevington hit a groundball to the shortstop who threw to first for the out. Dempsey followed with a single, but was out stealing when the catcher threw to the shortstop covering second base. Lubin began the third inning by popping out to the first baseman, then Gamboli doubled and scored on Wilson's single. O'Brien reached first base on an error by the third baseman and Wilson got to second on the same error. Chavez then hit into a double play. The shortstop fielded his grounder, threw to the second

baseman for one out, then the second baseman threw to first for the second out. O'Brien was out at second.

Now you're ready to score a game. Use the scoresheet below. You'll need a separate sheet for each team. Score a game that you listen to on the radio. Announcers know that many listeners keep score and they describe the action with that in mind.

Challenge problem

You can total each player's performance in the columns on the right of the card. Find out how many official at bats and hits one player had before the game, then combine what the player did in this game with those figures. You'll be able to compute his new batting average before it appears in tomorrow's newspaper! Make similar calculations for the teams' places in the standings.

PLAYER	POS.	1	2	3	4	5	6	7	8	9	10	11	Runs	Hits	RBI	A	PO	E
	2B																	
	LF																	
	3B																	
	1B																	
	C																	
	CF																	
	RF																	
	SS																	
	P																	
RUNS																		
HITS																		

TEAM

From Baseballmath: Grandslam Activities and Projects for Grades 4-8 published by GoodYearBooks. Copyright © 1994 Christopher Jennison.

Above average

Every Sunday during the baseball season the sports sections of most newspapers print player and team statistics. *USA Today* prints American League statistics every Tuesday, and National League statistics every Wednesday. Statistics for both leagues appear in *Baseball Weekly*. It's easy to see who the top players are and which teams are leading their divisions. But do the "star" players always play for the league-leading teams? Take a look at team batting averages, for instance. Are the best hitting teams leading their divisions? What about team earned run averages, the statistic used to measure a team's pitching strength? What about fielding averages? Perhaps some combination of these statistics would provide a better measure of a team's effectiveness. One simple computation would be to add a team's batting and fielding averages, divide by two, and subtract the team earned run average. Do this calculation for several teams. Does this "new" statistic provide a more accurate measure of a team's ability?

Using the calculation suggested above and the chart below, follow the progress of these teams over a period of time. You could do this for any time period you prefer, but the longer you follow the teams, the more reliable the information will be.

TEAMS	WEEK 1 Combined Average	WEEK 1 Division Standing	WEEK 2 Combined Average	WEEK 2 Division Standing	WEEK 3 Combined Average	WEEK 3 Division Standing	WEEK 4 Combined Average	WEEK 4 Division Standing	WEEK 5 Combined Average	WEEK 5 Division Standing	WEEK 6 Combined Average	WEEK 6 Division Standing

Was there any relationship between a team's combined statistics and their position in the division standings?

Can you name other factors that determine a team's success?

From *Baseballmath: Grandslam Activities and Projects for Grades 4-8* published by GoodYearBooks. Copyright © 1994 Christopher Jennison.

Home field

On page 70 is a diagram of a baseball stadium, surrounded by streets and city blocks. Many ballparks are located in inner-city areas, and their designs were carefully created to make maximum use of minimum space.

Now you have an opportunity to design your own ballpark, and add, if you wish, an imaginary surrounding neighborhood. Begin your design by drawing the baseball diamond and the foul lines. The foul lines extend from home plate at a 90° angle. The minimum distance from home plate to the foul pole in a major league stadium is 325 feet, but you can make it longer if you want to. (The batters on your team will not appreciate too long a distance, however.) The minimum distance from home plate to straight away centerfield is 400 feet, but this can be extended as well. Also, the minimum distance from home plate to the grandstand behind home plate is 60 feet. The distance between home plate and first base, and between all the other bases is 90 feet. Finally, the distance between home plate and the middle of the pitcher's mound is 60 feet, 6 inches.

Obviously your drawing can't show the actual distances, so you will have to use a scale. As a suggestion, use 1 inch for every 90 feet. That will simplify the measurements between bases. How many inches would you use if you wanted the distance from home plate to each foul pole to be 360 feet?

_____

How many inches would you use if you wanted the distance from home plate to straight-away center field to be 405 feet?

_____

After you have drawn the diamond and the foul lines, you can draw the grandstand and the outfield fences. The outfield fences can be curved, or shaped in almost any angle you can imagine, within reason. Make sure the minimum distances from home plate to the foul poles and center field are followed as described above. The grandstand can also be any shape you want it to be. The drawing on page 70 is only a suggestion. You can add as many details as you like, including dugouts, bullpens, seats, and lights.

To make your design even more interesting, you can draw streets and city blocks such as those shown on the drawing. You might want to include a parking lot near your ballpark. Details will make your drawing more interesting. Think about including subway stops, names of buildings, and names of streets. And think about a name for your ballpark. You might want to name it after yourself; after all, you did all the work designing it!

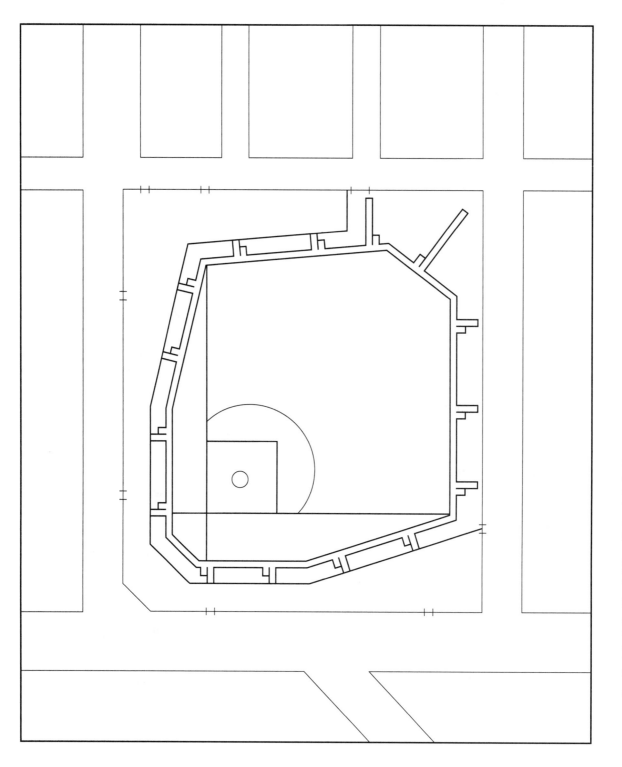

From *Baseballmath: Grandslam Activities and Projects for Grades 4-8* published by GoodYearBooks. Copyright © 1994 Christopher Jennison.

Lopsided ballpark

When the Brooklyn Dodgers left Brooklyn and became the Los Angeles Dodgers in 1958, they had to play for four seasons in the Los Angeles Coliseum before their new ballpark was built. The Coliseum was a perfect shape for football games and track meets, which it was designed for, but it was a poor place to play baseball. Take a look at the diagram below. It shows the distance from home plate to the foul poles and the outfield fence. As you can see, the distances to the foul poles are very short; in fact, the distance to the left field foul pole is the minimum distance permitted by the rules. A wire mesh screen, 42 feet high, was put up in front of the left field grandstand from the foul pole to a point 320 feet from home plate. This was done to keep pop flys from becoming homers, but in the first nine games played in the Coliseum 20 homers were hit over the screen. And although the right field foul pole is close to home plate, the right field fence actually started about 370 feet from home, which made it very hard for left-handed hitters to hit homers.

Maybe you can do a better job designing the playing field. The dimensions of the Coliseum will stay the same, of course, but could you reposition the playing area so that the distances down the foul lines are more identical? Begin by making a tracing of the Coliseum dimensions. Use the same scale (1 in. = 90 ft.) for drawing the diamond, and then see how far out the foul lines extend until they meet the grandstand. Are they at least the minimum distance of 250 feet in each case?

Do you think this is a better arrangement? Why?

Would you put up screens in front of the grandstands? Why or why not?

What distances would you set for the outfield fence? Why?

Why do you think the owners of the Dodgers arranged the playing field the way they did?

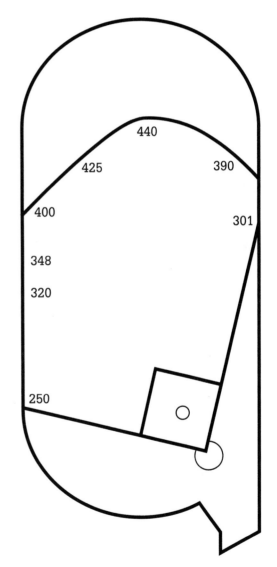

Star wars

It is fun to compare old-time players with players from recent years and try to figure out which ones were better. This project will not decide the issue, but it will give you a chance to make some comparisons and may enable you to strengthen your argument.

On the following eight pages are the names and career records of all players admitted to baseball's Hall of Fame through 1993. With two or three friends, select two all-star teams. One team should be made up of players born before 1925. The other team should include players born from 1925 and beyond. Select twenty-five players for each team. Ten players on each team should be pitchers.

Now select starting teams from each roster. For your starting teams compare the following:

1. A composite batting average for all players except the pitchers. (This is the average of the player's batting averages.)
2. The yearly home run averages for all players except pitchers.
3. The yearly RBI averages for all players except pitchers.
4. The yearly runs scored averages for all players except pitchers.
5. The yearly average of games won by each pitcher.
6. A composite earned run average for each pitcher.
7. A composite won-lost percentage for each pitcher.
8. The yearly average of shutouts by each pitcher.

Since baseball always involves more players than those on the starting team, make the same comparisons for the rest of the players on your squads. Based on these comparisons, do you want to make any changes in your starting lineups?

_____

Which starting lineup had the better statistics?

_____

Which overall team had the better statistics?

_____

To make the comparison, you can come up with team averages in batting and pitching.

Now you have some basis for comparing old-timers with more modern players. Are the modern players better?

Can you think of any other points to compare in deciding which players are better? Was baseball different during the early years of the century? In what way was it different, and how did it affect the players' performances?

From *Baseballmath: Grandslam Activities and Projects for Grades 4-8* published by GoodYearBooks. Copyright © 1994 Christopher Jennison.

To study these players and the years in which they played, you can read any one of a number of baseball history books in your school or local library. Some good ones include:

Alexander, Charles C. *Our Game: An American History.* New York: Holt, 1991.

The Baseball Encyclopedia: The Complete and Official Record of Major League Baseball, 9th ed. New York: Macmillan, 1993.

Okrent, Daniel, ed. *The Ultimate Baseball Book.* Rev. ed. New York: Houghton Mifflin, 1988.

Spalding, Daniel, ed. *America's National Game.* Omaha: University of Nebraska Press, 1992.

Thorn, John, & Palmer, Pete. *Total Baseball,* 3rd ed. New York: HarperCollins, 1993.

FIRST BASEMEN	YRS	B-T	HT.	WT.	BIRTHDATE	G	AB	R	H	2B	3B	HR	RBI	PCT
Anson, Cap	22	R-R	6-1	227	April 17, 1852	2253	9084	1712	2995	530	129	92	—	.339
Beckley, Jake	20	L-L	6-1	180	Aug. 4, 1867	2373	9476	1601	2930	455	46	87	1575	.309
Bottomley, Jim	16	L-L	6-0	180	April 23, 1900	1991	7471	1177	2313	465	151	219	1422	.310
Brouthers, Dan	19	L-L	6-2	207	May 8, 1858	1658	6725	1507	2349	446	212	103	—	.349
Chance, Frank	17	R-R	6-0	190	Sept. 9, 1877	1232	4279	796	1273	195	80	20	405	.297
Connor, Roger	18	L-L	6-2	210	July 1, 1857	1987	7807	1607	2535	429	227	132	—	.325
Foxx, Jimmie	20	R-R	6-0	190	Oct. 22, 1907	2317	8134	1751	2646	458	125	534	1921	.325
Gehrig, Lou	17	L-L	6-1	212	June 19, 1903	2164	8001	1888	2721	534	163	493	1990	.340
Greenberg, Hank	13	R-R	6-4	218	Jan. 1, 1911	1394	5193	1051	1628	379	71	331	1276	.313
Kelly, George	16	R-R	6-3	190	Sept. 10, 1896	1622	5993	8191	1778	337	76	148	1019	.297
Killebrew, Harmon	22	R-R	6-0	210	June 29, 1936	2435	8147	1283	2086	290	24	573	1584	.256
McCovey, Willie	22	L-L	6-4	225	Jan. 10, 1938	2588	8197	1229	2211	353	46	521	1555	.270
Mize, Johnny	15	L-R	6-2	215	Jan. 7, 1913	1884	6443	1118	2011	367	83	359	1337	.312
Sisler, George	15	L-L	5-10 1/2	170	Mar. 24, 1893	2055	8267	1284	2812	425	164	102	1180	.340
Stargell, Willie	21	L-L	6-3	220	Mar. 6, 1941	2360	7927	1195	2232	423	55	475	1540	.282
Terry, Bill	14	L-L	6-1/2	200	Oct. 30, 1898	1721	6428	1120	2193	373	112	154	1078	.341

SECOND BASEMEN	YRS	B-T	HT.	WT.	BIRTHDATE	G	AB	R	H	2B	3B	HR	RBI	PCT
Carew, Rod	19	L-R	6-0	182	Oct. 1, 1945	2469	9315	1424	3053	445	112	92	1015	.328
Collins, Eddie	25	L-R	5-9	175	May 2, 1887	2826	9946	1816	3309	437	186	47	1307	.333
Doerr, Bobby	14	R-R	5-11	185	April 7, 1918	1865	7093	1094	2042	381	89	223	1247	.288
Evers, Johnny	18	L-R	5-9	140	July 21, 1881	1776	6136	919	1659	216	70	12	538	.270
Frisch, Frankie	19	Bo-R	5-10	185	Sept. 9, 1989	2311	9112	1532	2880	466	138	105	1242	.316
Gehringer, Charlie	19	L-R	5-11 1/2	185	May 11, 1903	2323	8860	1774	2839	574	146	184	1427	.320
Herman, Billy	15	R-R	5-11	195	July 7, 1909	1922	7707	1163	2345	486	82	47	839	.304
Hornsby, Rogers	23	R-R	5-11	200	April 27, 1896	2259	8173	1579	2930	541	169	301	1579	.358
Lajoie, Nap	21	R-R	6-1	195	Sept. 5, 1875	2475	9589	1506	3252	652	164	82	1599	.339
Morgan, Joe	22	L-R	5-10	170	June 29, 1926	2649	9277	1650	2517	449	96	268	1133	.271
Robinson, Jackie	10	R-R	6-0	225	Jan. 31, 1919	1382	4877	947	1518	273	54	137	734	.311
Schoendienst, Al	19	Bo-R	6-0	170	Feb. 2, 1923	2216	8479	1223	2449	427	78	84	773	.289

SHORTSTOPS	YRS	a B-	HT.	WT.	BIRTHDATE	G	AB	R	H	2B	3B	HR	RBI	PCT
Aparicio, Luis	18	T	5-8	155	April 29, 1934	2599	10230	1335	2677	394	92	83	791	.262
Appling, Luke	20	R-R	5-11	200	April 2, 1907	2422	8856	1319	2749	440	102	45	1116	.310
Bancroft, Dave	16	R-R	5-9	160	April 20, 1892	1913	7182	1048	2004	320	77	32	579	.279
Banks, Ernie	19	Bo-R	6-1	180	Jan. 31, 1931	2528	9421	1305	2583	407	90	512	1636	.274
Boudreau, Lou	15	R-R	5-11	193	July 17, 1917	1646	6029	861	1779	385	66	68	789	.295
Cronin, Joe	20	R-R	5-11 1/2	180	Oct. 12, 1906	2124	7579	1233	2285	516	118	170	1423	.301
Jackson, Travis	15	R-R	5-10 1/2	160	Nov. 2, 1903	1656	6086	833	1768	291	86	135	929	.291
Jennings, Hugh	17	R-R	5-8 1/2	165	April 2, 1870	1264	4840	989	1520	227	88	19	840	.314
Maranville, Rabbit	23	R-R	5-5	155	Nov. 11, 1891	2670	10078	1255	2605	380	177	28	874	.258
Reese, Pee Wee	16	R-R	5-9 1/2	178	July 23, 1918	2166	8058	1338	2170	330	80	126	885	.269
Sewell, Joe	14	R-R	5-7	160	Oct. 9, 1989	1903	7132	1141	2226	436	68	49	1053	.312
Tinker, Joe	15	L-R	5-9	175	July 27, 1880	1642	5936	716	1565	238	106	29	—	.264
Vaughan, Arky	14	R-R	5-11	185	March 9, 1912	1817	6622	1173	2103	356	128	96	926	.318
Wagner, Honus	21	L-R	5-11	200	Feb. 24, 1874	2785	10427	1740	3430	651	252	101	1732	.329
Wallace, Bobby	25	R-R	5-8	170	Nov. 4, 1874	2369	8629	1056	2308	394	149	36	1121	.267
Ward, Monte	17	R-R	5-9	165	March 3, 1860	1810	7579	1403	2151	232	95	26	605	.283

From *Baseballmath: Grandslam Activities and Projects for Grades 4-8* published by GoodYearBooks. Copyright © 1994 Christopher Jennison.

THIRD BASEMEN	YRS	B-T	aHT.	WT.	BIRTHDATE	G	AB	R	H	2B	3B	HR	RBI	PCT
Baker, Frank	13	L-R	5-11 1/2	180	Mar. 13, 1886	1575	5983	887	1838	313	103	96	1012	.307
Collins, Jimmy	14	R-R	5-8	160	Jan. 16, 1873	1718	6792	1057	1999	333	117	62	985	.294
Kell, George	15	R-R	5-10	170	Aug. 23, 1922	1795	6702	881	2054	385	50	78	870	.306
Lindstrom, Fred	13	R-R	5-11 1/2	170	Nov. 21, 1905	1438	5611	895	1747	301	81	103	779	.311
Mathews, Eddie	17	L-R	6-1	200	Oct. 13, 1931	2391	8537	1509	2315	354	72	512	1453	.271
Robinson, Brooks	23	R-R	6-1	190	May 18, 1937	2896	10654	1232	2848	482	68	268	1357	.267
Traynor, Pie	17	R-R	6-1	175	Nov. 11, 1899	1941	7559	1183	2416	371	164	58	1273	.320

LEFT FIELDERS	YRS	B-T	HT.	WT.	BIRTHDATE	G	AB	R	H	2B	3B	HR	RBI	PCT
Brock, Lou	19	L-L	5-11 1/2	172	June 18, 1939	2616	10332	1610	3023	486	141	149	900	.293
Burkett, Jesse	16	L-L	5-8	155	Feb. 12, 1870	2063	8389	1708	2872	314	185	72	952	.342
Clarke, Fred	21	L-R	5-10 1/2	165	Oct. 3, 1872	2204	8584	1620	2703	358	219	65	1015	.315
Delahanty, Ed	16	R-R	5-10	170	Oct. 31, 1867	1825	7493	1596	2593	508	182	98	1464	.346
Goslin, Goose	18	L-R	5-11	180	Oct. 16, 1900	2287	8656	1483	2735	500	173	248	1609	.316
Hafey, Chick	13	R-R	6-1	185	Feb. 12, 1904	1283	4625	777	1466	341	67	164	833	.317
Kelley, Joe	17	R-R	5-11	190	Dec. 9, 1871	1829	6989	1425	2245	353	189	66	1194	.321
Kiner, Ralph	10	R-R	6-2	195	Oct. 27, 1922	1472	5205	971	1451	216	39	369	1015	.279
Manush, Heinie	17	L-L	6-0	200	July 20, 1901	2008	7654	1287	2524	491	160	110	1183	.330
Medwick, Joe	17	R-R	5-10	185	Nov. 4, 1911	1984	7635	1198	2471	540	113	205	1383	.324
Musial, Stan	22	L-L	6-0	180	Nov. 21, 1920	3026	10972	1949	3630	725	177	475	1951	.331
O'Rourke, Jim	19	R-R	5-8	185	Aug. 24, 1852	1750	7365	1425	2314	385	139	49	--	.314
Simmons, Al	20	R-R	6-0	210	May 22, 1902	2215	8761	1507	2927	539	149	307	1827	.334
Wheat, Zack	19	L-L	5-10	170	May 23, 1888	2406	9106	1289	2884	476	172	132	1265	.317
Williams, Billy	18	L-R	6-1	175	June 15, 1938	2488	9350	1410	2711	434	88	426	1475	.290
Williams, Ted	19	L-R	6-4	198	Aug. 30, 1918	2292	7706	1798	2654	525	71	521	1839	.344
Yastrzemski, Carl	23	L-R	5-11	185	Aug. 22, 1939	3308	11988	1816	3419	646	59	452	1844	.285

CENTER FIELDERS	YRS	B-T	HT.	WT.	BIRTHDATE	G	AB	R	H	2B	3B	HR	RBI	PCT
Averill, Earl	13	L-R	5-9 1/2	170	May 21, 1902	1669	6352	1224	2019	401	128	238	1164	.318
Carey, Max	20	Bo-R	6-0	166	Jan. 11, 1890	2469	9363	1545	2665	419	159	69	797	.285
Cobb, Ty	24	L-R	6-1	175	Dec. 18, 1886	3033	11429	2245	4191	724	298	117	1960	.367
Combs, Earle	12	L-R	6-0	185	May 14, 1899	1455	5746	1186	1866	309	154	58	629	.325
DiMaggio, Joe	13	R-R	6-2	195	Nov. 25, 1914	1736	6821	1390	2214	389	131	361	1537	.325
Duffy, Hugh	17	R-R	5-7	168	Nov. 26, 1866	1722	6999	1545	2307	310	117	103	597	.330
Hamilton, Billy	14	L-R	5-6	165	Feb. 16, 1866	1578	6262	1690	2157	225	94	37	736	.344
Mantle, Mickey	18	Bo-R	6-0	200	Oct. 20, 1931	2401	8102	1677	2415	344	72	536	1509	.298
Mays, Willie	22	R-R	5-11	187	May 6, 1931	2992	10881	2062	3283	523	140	660	1903	.302
Roush, Edd	18	L-L	5-11	175	May 8, 1893	1748	6646	1000	2158	311	168	63	882	.325
Snider, Duke	18	L-R	6-0	200	Sept. 19, 1926	2143	7161	1259	2116	358	85	407	1333	.295
Speaker, Tris	22	L-L	5-11 1/2	193	April 4, 1888	2791	10200	1881	3516	793	222	117	1562	.345
Waner, Lloyd	18	L-R	5-9	145	Mar. 16, 1906	1993	7772	1201	2459	281	118	27	598	.316
Wilson, Hack	12	R-R	5-6	195	April 26, 1900	1348	4760	884	1461	266	67	244	1062	.307

From *Baseballmath: Grandslam Activities and Projects for Grades 4-8* published by GoodYearBooks. Copyright © 1994 Christopher Jennison.

RIGHT FIELDERS	YRS	B-T	HT.	WT.	BIRTHDATE	G	AB	R	H	2B	3B	HR	RBI	PCT
Aaron, Hank	23	R-R	6-0	190	Feb. 5, 1934	3298	12364	2174	3771	624	98	755	2297	.305
Clemente, Roberto	18	R-R	5-11	182	Aug. 18, 1934	2433	9454	1416	3000	440	166	240	1305	.317
Crawford, Sam	19	L-L	6-0	190	April 18, 1880	2505	9579	1392	2964	455	312	95	1525	.309
Cuyler, Kiki	18	R-R	5-10 1/2	180	Aug. 30, 1899	1879	7161	1305	2299	394	157	128	1065	.321
Flick, Elmer	13	L-R	5-9	168	Jan. 11, 1876	1480	5601	951	1767	268	170	46	756	.315
Heilmann, Harry	17	R-R	6-1	200	Aug. 3, 1894	2145	7787	1291	2660	542	151	183	1552	.342
Hooper, Harry	17	L-R	5-10	165	Aug. 24, 1887	2308	8784	1429	2466	389	160	75	813	.281
Jackson, Reggie	21	L-L	6-0	200	April 18, 1946	2820	9864	1551	2584	463	49	563	1702	.262
Kaline, Al	22	R-R	6-2	184	Dec. 19, 1934	2834	10116	1622	3007	498	75	399	1583	.297
Keeler, Willie	19	L-L	5-4 1/2	140	Mar. 13, 1872	2124	8564	1720	2955	234	155	32	810	.345
Kelly, King	16	R-R	5-11	180	Dec. 31, 1857	1434	5922	1359	1853	351	109	65	--	.313
Klein, Chuck	17	L-R	6-0	195	Oct. 7, 1905	1753	6486	1168	2076	398	74	300	1201	.320
McCarthy, Tommy	13	R-R	5-7	170	July 24, 1864	1258	5055	1050	1485	195	57	43	--	.294
Ott, Mel	22	L-R	5-9	165	March 2, 1909	2730	9456	1859	2876	488	72	511	1860	.304
Rice, Sam	20	L-L	5-10	150	Feb. 20, 1890	2404	9269	1514	2987	498	184	34	1077	.322
Robinson, Frank	21	R-R	6-1	194	Aug. 31, 1935	2808	10006	1829	2943	528	72	586	1812	.294
Ruth, Babe	22	L-L	6-2	215	Feb. 6, 1895	2503	8397	2174	2873	506	136	714	2204	.342
Slaughter, Enos	19	L-R	5-9	190	April 27, 1916	2380	7946	1247	2383	413	148	169	1304	.300
Thompson, Sam	15	L-L	6-2	207	March 5, 1860	1405	6004	1259	2016	326	146	126	1299	.336
Waner, Paul	20	L-L	5-8	148	April 16, 1903	2549	9459	1627	3152	605	191	113	1309	.333
Youngs, Ross	10	Bo-R	5-8	162	April 10, 1897	1211	4627	812	1491	236	93	42	596	.322

From *Baseballmath: Grandslam Activities and Projects for Grades 4-8* published by GoodYearBooks. Copyright © 1994 Christopher Jennison.

CATCHERS	YRS	B-T	HT.	WT.	BIRTHDATE	G	AB	R	H	2B	3B	HR	RBI	PCT
Bench, Johnny	17	R-R	6-1	210	Dec. 7, 1947	2158	7658	1091	2048	381	24	389	1376	.267
Berra, Yogi	19	L-R	5-7 1/2	190	May 12, 1925	2120	7555	1175	2150	321	49	358	1430	.285
Bresnahan, Roger	17	R-R	5-8	180	June 11, 1879	1410	4480	684	1251	222	72	26	531	.279
Campanella, Roy	10	R-R	5-9 1/2	205	Nov. 19, 1921	1215	4205	627	1161	178	18	242	856	.276
Cochrane, Mickey	13	L-R	5-10 1/2	180	April 6, 1903	1482	5169	1041	1652	333	64	119	832	.320
Dickey, Bill	17	L-R	6-1 1/2	185	June 6, 1907	1789	6300	930	1969	343	72	202	1209	.313
Ewing, Buck	18	R-R	5-10	188	Oct. 27, 1859	1281	5348	1118	1663	237	179	66	--	.313
Ferrell, Rick	18	R-R	5-11	170	Oct. 12, 1905	1884	6028	687	1692	324	45	28	734	.28?
Hartnett, Gabby	20	R-R	6-1	218	Dec. 20, 1900	1990	6432	867	1912	396	64	236	1179	.29?
Lombardi, Ernie	17	R-R	6-3	230	April 6, 1908	1853	5855	601	1792	277	27	190	990	.30?
Schalk, Ray	18	R-R	5-7	154	Aug. 12, 1892	1760	5306	579	1345	199	48	12	596	.253

From *Baseballmath: Grandslam Activities and Projects for Grades 4-8* published by GoodYearBooks. Copyright © 1994 Christopher Jennison.

PITCHERS	YRS	B-T	HT.	WT.	BIRTHDATE	G	IP	SHO	W	L	PCT	H	SO	BB	ERA
Alexander, Grover	20	R-R	6-1	185	Feb. 26, 1887	696	5188	90	373	208	.642	4868	2198	951	2.56
Bender, Chief	16	R-R	6-2	185	May 5, 1884	433	2847	41	208	112	.650	2455	1630	667	--
Brown, Mordecai	14	Bo-R	5-10	175	Oct. 19, 1876	411	2697	55	208	111	.652	2284	1166	548	--
Chesbro, Jack	11	R-R	5-9	180	June 5, 1874	392	2886	35	198	128	.607	2602	1276	674	--
Clarkson, John	12	R-R	5-10	160	July 1, 1861	519	4514	37	327	176	.650	4384	2013	1192	--
Coveleski, Stan	14	R-R	5-9 1/2	175	July 13, 1890	450	3081	38	215	141	.604	3055	981	802	2.88
Dean, Dizzy	12	R-R	6-3	202	Jan. 16, 1911	317	1966	27	150	83	.644	1921	1155	458	3.04
Drysdale, Don	14	R-R	6-6	208	July 23, 1936	518	3432	49	209	166	.557	3084	2486	855	2.95
Faber, Red	20	Bo-R	6-1	190	Sept. 6, 1888	669	4089	29	254	212	.545	4104	1471	1213	3.15
Feller, Bob	18	R-R	6-0	185	Nov. 3, 1918	570	3828	46	266	162	.621	3271	2581	1764	3.25
Fingers, Rollie	17	R-R	6-4	195	Aug. 25, 1946	944	1701	2	114	118	.491	1474	1299	492	2.90
Ford, Whitey	16	L-L	5-10	180	Oct. 21, 1928	498	3171	45	236	106	.690	2766	1956	1086	2.74
Galvin, Pud	14	R-R	5-8	190	Dec. 25, 1856	697	5959	57	361	309	.539	6334	1786	744	--
Gibson, Bob	17	R-R	6-1	193	Nov. 9, 1935	528	3885	56	251	174	.591	3279	3117	1336	2.91
Gomez, Lefty	14	L-L	6-2	175	Nov. 26, 1908	368	2503	28	189	102	.649	2290	1468	1095	3.34
Grimes, Burleigh	19	R-R	5-10	195	Aug. 18, 1893	615	4178	35	270	212	.560	4406	1512	1295	3.52
Grove, Lefty	17	L-L	6-3	204	March 6, 1900	616	3940	35	300	141	.680	3849	2266	1187	3.06
Haines, Jesse	19	R-R	6-0	180	July 22, 1893	555	3208	23	210	158	.571	3460	981	871	3.64
Hoyt, Waite	21	R-R	5-11	183	Sept. 9, 1899	675	3762	26	237	182	.566	4037	1206	1003	3.59
Hubbell, Carl	16	R-L	6-1	172	June 22, 1903	535	3591	36	253	154	.622	3461	1677	725	2.98
Hunter, Catfish	15	R-R	6-0	190	April 18, 1946	500	3448	42	224	166	.574	2958	2012	954	3.26
Jenkins, Ferguson	19	R-R	6-5	210	Dec. 14, 1943	664	4500	49	284	226	.557	4142	3192	997	3.34
Johnson, Walter	21	R-R	6-1	200	Nov. 6, 1887	802	5923	113	416	279	.599	4920	3508	1353	--
Joss, Addie	9	R-R	6-3	185	April 12, 1880	286	2329	45	160	97	.623	1893	915	357	--
Keefe, Tim	14	R-R	5-10 1/2	185	Jan. 1, 1857	598	5043	40	342	224	.604	4524	2538	1225	--
Koufax, Sandy	12	R-L	6-2	202	Dec. 30, 1935	397	2325	40	165	87	.655	1754	2396	817	2.76
Lemon, Bob	13	L-R	6-0	180	Sept. 22, 1920	460	2849	31	207	128	.618	2559	1277	1251	3.23
Lyons, Ted	21	Bo-R	5-11	200	Dec. 28, 1900	594	4162	27	260	230	.531	4489	1073	1121	3.67
Marichal, Juan	16	R-R	5-11	190	Oct. 20, 1938	471	3506	52	243	142	.631	3153	2303	709	2.89
Marquard, Rube	18	Bo-L	6-3 1/2	175	Oct. 9, 1889	536	3307	30	201	177	.532	3233	1593	858	3.13

PITCHERS	YRS	B-T	HT.	WT.	BIRTHDATE	G	IP	SHO	W	L	PCT	H	SO	BB	ERA
Mathewson, Christy	17	R-R	6-1 1/2	195	Aug. 12, 1870	635	4781	83	373	188	.665	4203	2505	837	--
McGinnity, Joe	10	R-R	5-11	206	Mar.19, 1871	467	3455	32	247	145	.630	3235	1066	799	--
Nichols, Kid	15	R-R	5-10 1/2	175	Sept. 14, 1869	620	5067	48	361	208	.634	4854	1866	1245	--
Palmer, Jim	19	R-R	6-3	196	Oct. 15, 1945	558	3948	53	268	152	.638	3349	2212	1311	2.86
Pennock, Herb	22	R-R	6-0	165	Feb. 10, 1894	617	3572	35	240	162	.597	3900	1227	916	3.11
Perry, Gaylord	22	Bo-L	6-4	215	Sept. 15, 1938	777	5350	53	314	265	.542	4938	3534	1379	3.54
Plank, Eddie	17	L-L	5-11 1/2	175	Aug. 31, 1875	581	4234	69	305	181	.628	3688	2112	984	--
Radbourn, Old Hoss	11	R-R	5-9	168	Dec. 9, 1853	517	4543	35	308	191	.617	4500	1746	856	--
Rixey, Eppa	21	R-L	6-5 1/2	210	May 3, 1891	692	4494	39	266	251	.515	4633	1350	1350	3.15
Roberts, Robin	19	Bo-R	6-1	200	Sept. 30, 1926	676	4689	45	286	245	.539	4582	2357	2357	3.40
Ruffing, Red	22	R-R	6-1 1/2	215	May 3, 1905	624	4342	46	273	225	.548	4294	1987	1987	3.80
Rusie, Amos	10	R-R	6-1	210	May 31, 1871	412	3772	31	241	158	.604	3177	1953	1953	--
Seaver, Tom	20	R-R	6-1	206	Nov. 17, 1944	656	4782	61	311	205	.603	3971	3640	3640	2.86
Spahn, Warren	21	L-L	6-0	185	April 23, 1921	750	5246	63	363	245	.597	4830	2583	2583	3.08
Vance, Dazzy	16	R-R	6-1	200	March 4, 1891	442	2967	31	197	140	.585	2809	2045	2045	3.24
Waddell, Rube	13	L-L	6-1 1/2	196	Oct. 13, 1876	407	2958	50	191	142	.574	2480	2310	2310	--
Walsh, Ed	14	R-R	6-1	193	May 14, 1881	432	2969	58	195	126	.607	2335	1731	1731	--
Welch, Mickey	13	R-R	5-8	160	July 4, 1859	566	4794	41	307	209	.595	4646	1841	1841	--
Wilhelm, Hoyt	21	R-R	6-0	190	July 26, 1923	1070	2254	5	143	122	.540	1757	1610	1610	2.52
Wynn, Early	23	Bo-R	6-0	235	Jan. 6, 1920	691	4566	49	300	244	.551	4291	2334	2334	3.54
Young, Cy	22	R-R	6-2	210	Mar. 29, 1867	906	7377	77	511	313	.620	7078	2819	2819	--

Team yearbook

Most major-league teams publish yearbooks at the beginning of each season. The yearbooks include pictures of each player, their season and career records, some biographical information, and a summary of the previous season.

Now you can create your own team yearbook for your Little League team, or for any team that you may be part of. Begin by taking pictures of each of your teammates. Remember to ask someone to take a picture of you! You can use a Polaroid camera, or a regular camera. Try to vary the pictures; that is, use different poses so your book will have some variety. Some pictures can be head-and-shoulder shots, and others can be action pictures, or posed action pictures. You can get good ideas from your favorite major-league team's yearbook or any baseball magazine.

Once you have all the pictures developed, you can plan the layout of your yearbook. As a suggestion, use 8 1/2 x 11 sheets. These are very easy to find in drug stores or school supply stores. Measure how much space you want to reserve for each picture or pictures on one page, making sure to allow enough space for whatever text you want to provide. You

should write or type the text on the page before you paste on the pictures. Don't forget a page with all the names and ages of your teammates. You might even want to add their heights and weights, favorite major-league players, favorite teams, and so on. You can be as creative as you wish.

You'll want to have copies of the yearbook made for your teammates and others. Take your sheets to a copy shop, or ask someone to make photocopies for you. Then you can bind the yearbooks with a stapler, or punch three holes near the edge and place the yearbook in a three-hole portfolio. These can also be purchased at school supply stores or drugstores.

Think about asking a local businessperson to sponsor your yearbook. In turn, for helping you pay for the expenses of the yearbook, you could print a large advertisement of the business. Taking this one step further, think about placing lots of ads and distributing your yearbook to stores in your neighborhood or town. You might be able to raise enough money for new supplies for your team, or for keeping your field maintained. There are many possibilities. Local businesses are often very eager to help with projects like this. Good luck!

Lost seasons

It is fun to calculate what some players' final records might have been had they not lost playing time to injuries, or, as in the case of this project, to military service. Bob Feller and Ted Williams, two of baseball's most outstanding players, each lost three years in the prime of their careers to World War II. Williams lost most of two seasons to the Korean War, as well.

The career records for each player are reproduced on the student sheet, with questions for the students to answer. Some questions require library research for completion. Suggest to the students that they ignore the partial-season figures for Williams in 1952 and 1953, and make their calculations as if he did not play at all in those years. The final questions, in which the students are asked to guess whether the players might have done better or worse during the lost seasons, are obviously open-ended. Some students might guess that the players would have done better than average during the World War II years since those years came in the midst of their most productive seasons.

Answers: (These revised records are based on dividing the career totals by the number of years played. This computes an "average" season's productivity. Add the totals for each lost season to the current lifetime totals for new totals.)

Feller: Wins 314 (Tie for fourteenth on all-time list) Strikeouts 3016 (eleventh all-time) Winning Pct. .629 (all-time ranking not calculated)

Williams: Home runs 656 (fourth all-time) Hits 3354 (eighth all-time) Lifetime BA .344

From *Baseballmath: Grandslam Activities and Projects for Grades 4-8* published by GoodYearBooks. Copyright © 1994 Christopher Jennison.

On this and the next page are the career records of two of baseball's greatest players, Ted Williams and Bob Feller. But their records would have been even greater if they had not lost several seasons to military service. Both players were in the service for three years during World War II, and Williams also lost most of two seasons during the Korean War.

Williams, Theodore Samuel

Born, San Diego, California, August 30, 1918.
Bats Left. Throws Right. Height, 6 feet, 4 inches. Weight, 195 pounds.

Year	Club	Lea.	Pos.	G	AB	R	H	2B	3B	HR	RBI	SB	Avg.
1939	Boston	A. L.	OF	149	565	131	185	44	11	31	145	2	.327
1940	Boston	A. L.	OF-P	144	561	134	193	43	14	23	113	4	.344
1941	Boston	A. L.	OF	143	456	135	185	33	3	37	120	2	.406
1942	Boston	A. L.	OF	150	522	141	186	34	5	36	137	3	.356
1943-44-45	Boston	A. L.		(in Military Service)									
1946	Boston	A. L.	OF	150	514	142	176	37	8	38	123	0	.342
1947	Boston	A. L.	OF	156	528	125	181	40	9	32	114	0	.343
1948	Boston	A. L.	OF	137	509	124	188	44	3	25	127	4	.369
1949	Boston	A. L.	OF	155	566	150	194	39	3	43	159	1	.343
1950	Boston	A. L.	OF	89	334	82	106	24	1	28	97	3	.317
1951	Boston	A. L.	OF	148	531	109	169	28	4	30	126	1	.318
1952	Boston	A. L.	OF	6	10	2	4	0	1	1	3	0	.400
1953	Boston	A. L.	OF	37	91	17	37	6	0	13	34	0	.407
1954	Boston	A. L.	OF	117	386	93	133	23	1	29	89	0	.345
1955	Boston	A. L.	OF	98	320	77	114	21	3	28	83	2	.356
1956	Boston	A. L.	OF	136	400	71	138	28	2	24	82	0	.345
1957	Boston	A. L.	OF	132	420	96	163	28	1	38	87	0	.388
1958	Boston	A. L.	OF	129	411	81	135	23	2	26	85	1	.328
1959	Boston	A. L.	OF	103	272	32	69	15	0	10	43	0	.254
1960	Boston	A. L.	OF	113	310	56	98	15	0	29	72	1	.316
Major League Totals	19 Years			2292	7706	1798	2654	525	71	521	1839	24	.344

If he had not been called to military, how many games might Feller have won? _____

How many strikeouts would he have recorded? _____

What might have been his lifetime won-lost percentage? _____

If Williams had not lost so many years to the military, how many homeruns might he have hit? _____

What would have been his total number of hits? _____

What might have been his Lifetime batting average? _____

83 Lost seasons

From Baseballmath: Grandslam Activities and Projects for Grades 4-8 published by GoodYearBooks. Copyright © 1994 Christopher Jennison.

Feller, Robert William Andrew

Born, Van Meter, Iowa. November 3, 1918.

Bats Right. Throws Right. Height, 6 feet. Weight, 185 pounds.

Year	Club	Lea.	G	IP	W	L	Pct.	SO	BB	H	ERA
1936	Cleveland	A. L.	14	62	5	3	.625	76	47	52	3.34
1937	Cleveland	A. L.	26	149	9	7	.563	150	106	116	3.38
1938	Cleveland	A. L.	39	278	17	11	.607	240	208	225	4.08
1939	Cleveland	A. L.	39	297	24	9	.727	246	142	227	2.85
1940	Cleveland	A. L.	43	320	27	11	.711	261	118	245	2.62
1941	Cleveland	A. L.	44	343	25	13	.658	260	194	284	3.15
1942-43-44	Cleveland	A. L.			(In Military Service						
1945	Cleveland	A. L.	9	72	5	3	.625	59	35	50	2.50
1946	Cleveland	A. L.	48	371	26	15	.634	348	153	277	2.18
1947	Cleveland	A. L.	42	299	20	11	.645	196	127	230	2.68
1948	Cleveland	A. L.	44	280	19	15	.559	164	116	255	3.57
1949	Cleveland	A. L.	36	211	15	14	.517	108	84	198	3.75
1950	Cleveland	A. L.	35	247	16	11	.593	119	103	230	3.43
1951	Cleveland	A. L.	33	250	22	8	.733	111	95	239	3.49
1952	Cleveland	A. L.	30	192	9	13	.409	81	83	219	4.73
1953	Cleveland	A. L.	25	176	10	7	.588	60	60	163	3.58
1954	Cleveland	A. L.	19	140	13	3	.813	59	39	127	3.09
1955	Cleveland	A. L.	25	83	4	4	.500	25	31	71	3.47
Major League Totals 17 Years			551	3770	266	158	.627	2563	1741	3208	3.23

Look at your new figures for Williams's homers, hits, and Lifetime batting average. Where would he stand now among the all time leaders in these categories?

Homers _____

Hits _____

BA _____

(Your school or local Librarian will help you find reference books to use to find these statistics)

Look at Feller's new figures. Where would he stand among League Leaders Now?

Wins _____

Strikeouts _____

Won-Lost Pct. _____

Do you think the players might have done better or worse than average during the years they were in the service? Why?

From *Baseballmath: Grandslam Activities and Projects for Grades 4-8* published by GoodYearBooks. Copyright © 1994 Christopher Jennison.

Baseball's honor roll

Batting and pitching records are provided on pages 86–92. Lifetime records for homeruns, hits, batting average, and runs-batted-in are listed. Pitching records include wins, saves, strikeouts, and earned run average. Also listed are single-season records in the same categories. Some records are shown by era, since it is not always meaningful to compare today's game to the way baseball was played earlier in the century.

As you've discovered, baseball interest thrives on numbers and statistics. Some youngsters find endless fascination in comparing and predicting performances based on average and cumulative statistics. For example, in the lifetime homerun category there are three active players: Dave Winfield, Eddie Murray, and Andre Dawson. Based on these players' average number of homeruns per season, it's possible to estimate how many homeruns they will hit by the time they retire. You would need to estimate how many more years a player would be likely to play and also allow for some decrease in a player's seasonal total as he gets closer to retirement.

In the pitching categories, Nolan Ryan looks as if he will remain the all-time king of strikeout pitchers. Can anyone catch him? Only three active players appear on the list of career strikeout leaders. All are very close to retirement. One player not on this list is Roger Clemens. At the end of the 1993 season he had a total of 2,033 strikeouts. He has pitched in the majors for ten seasons. Based on an average number of strikeouts per season, how many more seasons would Clemens have to pitch for him to exceed Ryan's total? How old would he be? How would that compare with Ryan's when he retired?

These are just a few examples of problems and research activities that can be based on the game's abundant statistics and records.

Baseball's Honor Roll
Lifetime Pitching Records (Through 1993)

	Wins	
1	Cy Young	511
2	Walter Johnson	417
3	Pete Alexander	373
	Christy Mathewson	373
5	Warren Spahn	363
6	Kid Nichols	361
7	Jim Galvin	360
8	Tim Keefe	342
9	Steve Carlton	329
10	John Clarkson	328
11	Eddie Plank	326
12	Don Sutton	324
	Nolan Ryan	324
14	Phil Niekro	318
15	Gaylord Perry	314
16	Tom Seaver	311
17	Charley Radbourn	309
18	Mickey Welch	307
19	Lefty Grove	300
	Early Wynn	300
21	Tommy John	288
22	Bert Blyleven	287
23	Robin Roberts	286
24	Fergie Jenkins	284
	Tony Mullane	284
26	Jim Kaat	283
27	Red Ruffing	273
28	Burleigh Grimes	270
29	Jim Palmer	268
30	Bob Feller	266
	Eppa Rixey	266
32	Jim McCormick	265
33	Gus Weyhing	264
34	Ted Lyons	260
35	Red Faber	254
36	Carl Hubbell	253
37	Bob Gibson	251
38	Vic Willis	249
39	Jack Quinn	247
40	Joe McGinnity	246

	Strikeouts	
1	Nolan Ryan	5932
2	Steve Carlton	4136
3	Bert Blyleven	3701
4	Tom Seaver	3640
5	Don Sutton	3574
6	Gaylord Perry	3534
7	Walter Johnson	3509
8	Phil Niekro	3342
9	Fergie Jenkins	3192
10	Bob Gibson	3117
11	Jim Bunning	2855
12	Mickey Lolich	2832
13	Cy Young	2800
14	Frank Tanana	2773
15	Warren Spahn	2583
16	Bob Feller	2581
17	Jerry Koosman	2556
18	Tim Keefe	2545
19	Christy Mathewson	2502
20	Don Drysdale	2486
21	Jim Kaat	2461
22	Sam McDowell	2453
23	Luis Tiant	2416
24	Sandy Koufax	2396
25	Jack Morris	2378
26	Robin Roberts	2357
27	Early Wynn	2334
28	Rube Waddell	2316
29	Juan Marichal	2303
30	Lefty Grove	2266
31	Eddie Plank	2246
32	Tommy John	2245
33	Jim Palmer	2212
34	Pete Alexander	2198
35	Dennis Eckersley	2198
36	Vida Blue	2175
37	Charlie Hough	2171
38	Camilo Pascual	2167
39	Bobo Newsome	2082
40	Dazzy Vance	2045

From *Baseballmath: Grandslam Activities and Projects for Grades 4-8* published by GoodYearBooks. Copyright © 1994 Christopher Jennison.

	Saves	
1	Lee Smith	358
2	Jeff Reardon	357
3	Rollie Fingers	341
4	Rich Gossage	309
5	Bruce Sutter	300
6	Dennis Eckersley	275
7	Tom Henke	260
8	Dave Righetti	252
9	Dan Quisenberry	244
10	Sparky Lyle	238
11	John Franco	236
12	Hoyt Wilhelm	227
13	Gene Garber	218
14	Dave Smith	216
15	Bobby Thigpen	201
16	Roy Face	193
17	Mike Marshall	188
18	Steve Bedrosian	184
	Kent Tekulve	184
20	Tug McGraw	180
21	Ron Perranoski	179
22	Lindy McDaniel	172
23	Doug Jones	164
24	Stu Miller	154
25	Jay Howell	153
26	Don McMahon	153

Earned Run Average (By Era)

1942-1960

1	Hoyt Wilhelm	2.52
2	Whitey Ford	2.75
3	Harry Brecheen	2.92
4	Mort Cooper	2.97
5	Max Lanier	3.01
6	Tiny Bonham	3.06
	Hal Newhouser	3.06
8	Warren Spahn	3.09
9	Sal Maglie	3.15
10	Ed Lopat	3.21
11	Bob Lemon	3.23
	Dizzy Trout	3.23
13	Stu Miller	3.24
14	Bob Feller	3.25
	Dutch Leonard	3.25

1961-1992

1	Sandy Koufax	2.76
2	Roger Clemens	2.80
3	Andy Messersmith	2.86
	Jim Palmer	2.86
	Tom Seaver	2.86
6	Orel Hershiser	2.87
7	Juan Marichal	2.89
8	Rollie Fingers	2.90
9	Bob Gibson	2.91
10	Dean Chance	2.92
11	Rich Gossage	2.93
12	Don Drysdale	2.95
13	Mel Stottlemyre	2.97
14	Dwight Gooden	2.99
15	Bob Veale	3.07

Baseball's Honor Roll
Single Season Pitching Records

Wins (By Era)

1942-1960

1	Hal Newhouser, 1944	29
2	Robin Roberts, 1952	28
3	Dizzy Trout, 1944	27
	Don Newcombe, 1956	27
5	Hal Newhouser, 1946	26
	Bob Feller, 1946	26
7	Hal Newhouser, 1945	25
	Dave Ferriss, 1946	25
	Mel Parnell, 1949	25
10	Johnny Sain, 1948	24
11	Bobby Shantz, 1952	24
12	13 players tied	23

1961-1992

1	Denny McLain, 1968	31
2	Sandy Koufax, 1966	27
	Steve Carlton, 1972	27
	Bob Welch, 1990	27
5	Sandy Koufax, 1965	26
	Juan Marichal, 1968	26
7	12 players tied	25

Strikeouts (By Era)

1942-1960

1	Bob Feller, 1946	348
2	Hal Newhouser, 1946	275
3	Herb Score, 1956	263
4	Don Drysdale, 1960	246
5	Herb Score, 1955	245
6	Don Drysdale, 1959	242
7	Sam Jones, 1958	225
8	Hal Newhouser, 1945	212
9	Bob Turley, 1955	210
10	Sam Jones, 1959	209
11	Jim Bunning, 1959	201
	Jim Bunning, 1959	201
13	Robin Roberts, 1953	198
	Sam Jones, 1955	198
15	Sandy Koufax, 1960	197

1961-1992

1	Nolan Ryan, 1973	383
2	Sandy Koufax, 1965	382
3	Nolan Ryan, 1974	367
4	Nolan Ryan, 1977	341
5	Nolan Ryan, 1972	329
6	Nolan Ryan, 1976	327
7	Sam McDowell, 1965	325
8	Sandy Koufax, 1966	317
9	J. R. Richard, 1979	313
10	Steve Carlton, 1972	310
11	Mickey Lolich, 1971	308
12	Sandy Koufax, 1963	306
	Mike Scott, 1986	306
14	Sam McDowell, 1970	304
15	J. R. Richard, 1978	303

From *Baseballmath: Grandslam Activities and Projects for Grades 4-8* published by GoodYearBooks. Copyright © 1994 Christopher Jennison.

Earned Run Average (By Era)

1942-1960

1	Spud Chandler, 1943	1.64
2	Mort Cooper, 1942	1.78
3	Hal Newhouser, 1945	1.81
4	Max Lanier, 1943	1.90
5	Hal Newhouser, 1946	1.94
6	Billy Pierce, 1955	1.97
7	Whitey Ford, 1958	2.01
8	Al Benton, 1945	2.02
9	Allie Reynolds, 1952	2.06
10	Ted Lyons, 1942	2.10
	Howie Pollet, 1946	2.10
	Spud Chandler, 1946	2.10
	Warren Spahn, 1953	2.10
14	Dizzy Trout, 1944	2.12
	Roger Wolff, 1945	2.12

1961-1992

1	Bob Gibson, 1968	1.12
2	Dwight Gooden, 1985	1.53
3	Luis Tiant, 1968	1.60
4	Dean Chance, 1964	1.65
5	Nolan Ryan, 1981	1.69
6	Sandy Koufax, 1966	1.73
7	Sandy Koufax, 1964	1.74
8	Ron Guidry, 1978	1.74
9	Tom Seaver, 1971	1.76
10	Sam McDowell, 1968	1.81
11	Vida Blue, 1971	1.82
12	Phil Niekro, 1967	1.87
	Joe Horlen, 1964	1.88
14	Sandy Koufax, 1963	1.88
15	Luis Tiant, 1972	1.91

Saves

1	Bobby Thigpen, 1990	57
2	Dennis Eckersley, 1992	51
3	Dennis Eckersley, 1990	48
4	Lee Smith, 1991	47
5	Dave Righetti, 1986	46
	Bryan Harvey, 1991	46
7	Dan Quisenberry, 1983	45
	Bruce Sutter, 1984	45
	Dennis Eckersley, 1988	45
10	Dan Quisenberry, 1984	44
	Mark Davis, 1989	44
12	Doug Jones, 1990	43
	Dennis Eckersley, 1991	43
	Lee Smith, 1992	43
15	Jeff Reardon, 1988	42
	Rick Aguilera, 1991	42
17	Jeff Reardon, 1985	41
	Rick Aguilera, 1992	41
19	Steve Bedrosian, 1987	40
	Jeff Reardon, 1991	40
21	Johnny Franco, 1988	39
	Jeff Montgomery, 1992	39
23	John Hiller, 1973	38
	Jeff Russell, 1989	38
	Randy Myers, 1992	38
26	Clay Carroll, 1972	37
	Rollie Fingers, 1978	37
	Bruce Sutter, 1979	37
	Dan Quisenberry, 1985	37
	Doug Jones, 1988	37
	Gregg Olson, 1990	37
	John Wetteland, 1992	37

Baseball's Honor Roll

Lifetime Batting Records (Through 1993)

	Home Runs	
1	Hank Aaron	755
2	Babe Ruth	714
3	Willie Mays	660
4	Frank Robinson	586
5	Harmon Killebrew	573
6	Reggie Jackson	563
7	Mike Schmidt	548
8	Mickey Mantle	536
9	Jimmie Foxx	534
10	Willie McCovey	521
	Ted Williams	521
12	Ernie Banks	512
	Eddie Mathews	512
14	Mel Ott	511
15	Lou Gehrig	493
16	Stan Musial	475
	Willie Stargell	475
18	Dave Winfield	453
19	Carl Yastrzemski	452
20	Dave Kingman	442
21	Eddie Murray	441
22	Billy Williams	426
23	Darrell Evans	414
24	Andre Dawson	412
25	Duke Snider	407
26	Al Kaline	399
27	Dale Murphy	398
28	Graig Nettles	390
29	Johnny Bench	389
30	Dwight Evans	385
31	Frank Howard	382
	Jim Rice	382
33	Orlando Cepeda	379
	Tony Perez	379
35	Norm Cash	377
36	Carlton Fisk	376
37	Rocky Colavito	374
38	Gil Hodges	370
39	Ralph Kiner	369
40	Joe DiMaggio	361
41	Johnny Mize	359
42	Yogi Berra	358

	Batting Average	
1	Ty Cobb	.366
2	Rogers Hornsby	.358
3	Joe Jackson	.356
4	Ed Delahanty	.346
5	Tris Speaker	.345
6	Ted Williams	.344
	Billy Hamilton	.344
8	Dan Brouthers	.342
	Babe Ruth	.342
	Harry Hellmann	.342
11	Pete Browning	.341
	Willie Keeler	.341
	Bill Terry	.341
14	George Sisler	.340
	Lou Gehrig	.340
16	Jesse Burkett	.338
	Nap Lajoie	.338
18	Riggs Stephenson	.336
19	Wade Boggs	.335
20	Al Simmons	.334
	John McGraw	.334
22	Paul Waner	.333
	Eddie Collins	.333
	Mike Donlin	.333
25	Stan Musial	.331
	Sam Thompson	.331
27	Heinie Manush	.330
28	Cap Anson	.329
	Tony Gwynn	.329
30	Rod Carew	.328
31	Honus Wagner	.327
32	Tip O'Neill	.326
33	Bob Fothergill	.325
	Jimmie Foxx	.325
	Earle Combs	.325
	Joe DiMaggio	325

From *Baseballmath: Grandslam Activities and Projects for Grades 4-8* published by GoodYearBooks. Copyright © 1994 Christopher Jennison.

Hits

1	Pete Rose	4256
2	Ty Cobb	4189
3	Hank Aaron	3771
4	Stan Musial	3630
5	Tris Speaker	3514
6	Carl Yastrzemski	3419
7	Honus Wagner	3415
8	Eddie Collins	3312
9	Willie Mays	3283
10	Nap Lajoie	3242
11	George Brett	3154
12	Paul Waner	3152
13	Robin Yount	3142
14	Rod Carew	3053
15	Lou Brock	3023
16	Al Kaline	3007
17	Dave Winfield	3014
18	Roberto Clemente	3000
19	Cap Anson	2995
20	Sam Rice	2987
21	Sam Crawford	2961
22	Frank Robinson	2943
23	Willie Keeler	2932
24	Jake Beckley	2930
	Rogers Hornsby	2930
26	Al Simmons	2927
27	Zack Wheat	2884
28	Frankie Frisch	2880
29	Mel Ott	2876
30	Babe Ruth	2873
31	Jesse Burkett	2850
32	Brooks Robinson	2848
33	Charlie Gehringer	2839
34	George Sisler	2812
35	Vada Pinson	2757

Runs Batted In

1	Hank Aaron	2297
2	Babe Ruth	2213
3	Lou Gehrig	1995
4	Stan Musial	1951
5	Ty Cobb	1937
6	Jimmie Foxx	1922
7	Willie Mays	1903
8	Cap Anson	1879
9	Mel Ott	1860
10	Carl Yastrzemski	1844
11	Ted Williams	1839
12	Al Simmons	1827
13	Frank Robinson	1812
14	Dave Winfield	1786
15	Honus Wagner	1732
16	Reggie Jackson	1702
17	Eddie Murray	1662
18	Tony Perez	1652
19	Ernie Banks	1636
20	Goose Goslin	1609
21	Nap Lajoie	1599
22	Mike Schmidt	1595
	George Brett	1595
24	Rogers Hornsby	1584
	Harmon Killebrew	1584
26	Al Kaline	1583
27	Jake Beckley	1575
28	Willie McCovey	1555
29	Willie Stargell	1540
30	Harry Heilmann	1539
31	Joe DiMaggio	1537
32	Tris Speaker	1529
33	Sam Crawford	1525
34	Mickey Mantle	1509
35	Dave Parker	1493
35	Andre Dawson	1492
37	Billy Williams	1475
38	Rusty Staub	1466
39	Ed Delahanty	1464
40	Eddie Mathews	1453

Single Season Batting Records

Home Runs

1	Roger Maris, 1961	61
2	Babe Ruth, 1927	60
3	Babe Ruth, 1921	59
4	Jimmie Foxx, 1932	58
	Hank Greenberg, 1938	58
6	Hack Wilson, 1930	56
7	Babe Ruth, 1920	54
	Babe Ruth, 1928	54
	Ralph Kiner, 1949	54
	Mickey Mantle, 1961	54
11	Mickey Mantle, 1956	52
	Willie Mays, 1965	52
	George Foster, 1977	52
14	Ralph Kiner, 1947	51
	Johnny Mize, 1947	51
	Willie Mays, 1955	51
	Cecil Fielder, 1990	51
18	Jimmie Foxx, 1938	50
19	Babe Ruth, 1930	49
	Lou Gehrig, 1934	49
	Lou Gehrig, 1936	49
	Ted Kluszewski, 1954	49
	Willie Mays, 1962	49
	Harmon Killebrew, 1964	49
	Frank Robinson, 1966	49
	Andre Dawson, 1987	49
	Mark McGwire, 1987	49

Baseball's Honor Roll

Single Season Batting Records

Batting Average (By Era)

1942-1960

1	Ted Williams, 1957	.388
2	Stan Musial, 1948	.376
3	Ted Williams, 1948	.369
4	Stan Musial, 1946	.365
	Mickey Mantle, 1957	.365
6	Harry Walker, 1947	.363
7	Dixie Walker, 1944	.357
	Stan Musial, 1943	.357
9	Ted Williams, 1942	.356
10	Phil Cavaretta, 1945	.355
	Lou Boudreau, 1948	.355
	Stan Musial, 1951	.355
	Hank Aaron, 1959	.355
14	Billy Goodman, 1950	.354
15	Harvey Kuenn, 1959	.353

1961-1992

1	George Brett, 1980	.390
2	Rod Carew, 1977	.388
3	Tony Gwynn, 1987	.370
4	Wade Boggs, 1985	.368
5	Wade Boggs, 1988	.366
	Rico Carty, 1970	.366
7	Rod Carew, 1974	.364
8	Wade Boggs, 1987	.363
	Joe Torre, 1971	.363
10	Wade Boggs, 1983	.361
	Norm Cash, 1961	.361
12	Rod Carew, 1975	.359
13	Roberto Clemente, 1967	.357
	Wade Boggs, 1986	.357
15	Kirby Puckett, 1988	.356

Hits

1	George Sisler, 1920	257
2	Lefty O'Doul, 1929	254
	Bill Terry, 1930	254
4	Al Simmons, 1925	253
5	Rogers Hornsby, 1922	250
	Chuck Klein, 1930	250
7	Ty Cobb, 1911	248
8	George Sisler, 1922	246
9	Heinie Manush, 1928	241
	Babe Herman, 1930	241
11	Jesse Burkett, 1896	240
	Wade Boggs, 1985	240
13	Willie Keeler, 1897	239
	Rod Carew, 1977	239
15	Ed Delahanty, 1899	238
	Don Mattingly, 1986	238
17	Hugh Duffy, 1894	237
	Harry Heilmann, 1921	237
	Paul Waner, 1927	237
	Joe Medwick, 1937	237
21	Jack Tobin, 1921	236
22	Rogers Hornsby, 1921	235
23	Lloyd Waner, 1929	234
	Kirby Puckett, 1988	234

Runs Batted In

1	Hack Wilson, 1930	190
2	Lou Gehrig, 1931	184
3	Hank Greenberg, 1937	183
4	Lou Gehrig, 1927	175
	Jimmie Foxx, 1938	175
6	Lou Gehrig, 1930	174
7	Babe Ruth, 1921	171
8	Chuck Klein, 1930	170
	Hank Greenberg, 1935	170
10	Jimmie Foxx, 1932	169
11	Joe DiMaggio, 1937	167
12	Sam Thompson, 1887	166
13	Sam Thompson, 1895	165
	Al Simmons, 1930	165
	Lou Gehrig, 1934	165
16	Babe Ruth, 1927	164
17	Babe Ruth, 1931	163
	Jimmie Foxx, 1933	163
19	Hal Trosky, 1936	162
20	Hack Wilson, 1929	159
	Lou Gehrig, 1937	159
	Ted Williams, 1949	159
	Vern Stephens, 1949	159
24	Al Simmons , 1929	157
25	Jimmie Foxx, 1930	156
26	Ken Williams, 1922	155
	Joe DiMaggio, 1948	155
28	Babe Ruth, 1929	154
	Joe Medwick, 1937	154

From *Baseballmath: Grandslam Activities and Projects for Grades 4-8* published by GoodYearBooks. Copyright © 1994 Christopher Jennison.

"Casey at the Bat"

More than one hundred years ago, a young man named Ernest Lawrence Thayer published a poem titled "Casey at the Bat" in the pages of the *San Francisco Examiner* (June 3, 1888). It has been a favorite piece of American folklore ever since. It is great fun to read or listen to, but it's even more fun to perform. One person can recite the poem for the enjoyment of all, or you can involve your whole class by assigning parts to a few, and asking the rest of the class to play the part of the crowd.

Someone must play the narrator, and three other students must play the part of Casey, the pitcher, and the umpire. If you have enough room, other students can play the parts of the various fielders and the base runners. The poem is reprinted on the following page with notes in the margin to help you dramatize the action.

For follow-up activities, some students could write newspaper reports of the game. Other people could interview Casey and report his comments and feelings. And the questions below could be assigned individually, or discussed at the end of the dramatization.

1. What was the final score of the game, and in what inning does the action take place?

2. How many spectators were in the stands?

3. If Casey had 188 hits and 515 at bats before his last appearance, what was his batting average after he struck out?

4. Why were the two hitters preceding Casey in the lineup portrayed so hopelessly? A hitter of Casey's stature would normally bat third or fourth in the lineup, and the batters just ahead of him would be among the team's best hitters. What sort of manager would put two poor hitters up ahead of Casey? Why? Could Casey have been a pinch-hitter? Is there any evidence of this? What did the poet mean when he called Flynn a lulu and Blake a cake? Did these terms have different meanings a hundred years ago?

Finally, you could ask a few students to do some financial calculations. Assume that the Mudville team's expenses that day were $100,000 for salaries, maintenance, advertising, etc. After the students have remembered how many spectators were at the game, ask them to calculate how much the spectators would have to be charged for tickets in order for the team to make a profit of 20 percent. To make the project a little more interesting, imagine that 30 percent of the seats were box seats, 60 percent were grandstand seats, and 10 percent were bleacher seats. Grandstand seats should be priced at twice the rate of bleacher seats, and box seats at twice the rate of grandstand seats.

The outlook wasn't brilliant for the Mudville nine that day; *Narrator reads*

The score stood four to two with but one inning more to play.

And then when Cooney died at first, and Barrows did the same, *1) Pantomime two runners out at first.*

A sickly silence fell upon the patrons of the game. *2) Crowd sighs, then falls silent.*

A straggling few got up to go in deep despair. The rest *3) A few get up and leave in disgust.*

Clung to that hope which springs eternal in the human breast;

They thought, "If only Casey could get a whack at that—

We'd put up even money now with Casey at the bat."

But Flynn preceded Casey, as did also Jimmy Blake,

And the former was a lulu and the latter was a cake;

So upon that stricken multitude grim melancholy sat, *4) The crowd sighs again*

For there seemed but little chance of Casey's getting to the bat.

But Flynn let drive a single, to the wonderment of all,

And Blake, the much despised, tore the cover off the ball;

And when the dust had lifted, and the men saw what had

occurred, *5) Pantomime runners reaching second and third.*

There was Jimmy safe at second and Flynn a-hugging third.

From *Baseballmath: Grandslam Activities and Projects for Grades 4-8* published by GoodYearBooks. Copyright © 1994 Christopher Jennison.

 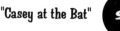

Then from five thousand throats and more there rose a lusty yell; *6) Crowd cheers.*

It rumbled through the valley, it rattled in the dell;

It knocked upon the mountain and recoiled upon the flat,

For Casey, mighty Casey, was advancing to the bat.

There was ease in Casey's manner as he stepped into his place; *7) Casey's gestures should be very extravagant.*

There was pride in Casey's bearing and a smile on Casey's face.

And when, responding to the cheers, he lightly doffed his hat,

No stranger in the crowd could doubt 'twas Casey at the bat.

Ten thousand eyes were on him as he rubbed his hands with dirt;

Five thousand tongues applauded when he wiped them on his

    shirt. *8) More cheers*

Then while the writhing pitcher ground the ball into his hip,

Defiance gleamed in Casey's eye, a sneer curled Casey's lip.

And now the leather-covered sphere came hurtling through the air, *9) More pantomime*

And Casey stood a-watching it in haughty grandeur there.

Close by the sturdy batsman the ball unheeded sped —

"That ain't my style," said Casey. "Strike one," the umpire said. *10) Casey and umpire speak*

From the benches, black with people, there went up a muffled

    roar, *11) Crowd reacts*

Like the beating of the storm-waves on a stern and distant shore;

"Kill him! Kill the umpire!" shouted some one on the stand;

And it's likely they'd have killed him had not Casey raised his

    hand.

"Casey at the Bat"

With a smile of Christian charity great Casey's visage shone;   *12) More grand gestures from Casey*

He stilled the rising tumult; he bade the game go on;

He signaled to the pitcher, and once more the spheroid flew;

But Casey still ignored it, and the umpire said, "Strike two."

"Fraud!" cried the maddened thousands, and echo answered

     "Fraud!"   *13) Crowd reacts*

But one scornful look from Casey and the audience was awed.

They saw his face grow stern and cold, they saw his muscles

    strain,

And they knew that Casey wouldn't let that ball go by again.

The sneer is gone from Casey's lip, his teeth are clenched in hate;

He pounds with cruel violence his bat upon the plate.   *14) More pantomime*

And now the pitcher holds the ball, and now he lets it go,

And now the air is shattered by the force of Casey's blow.

Oh, somewhere in this favored land the sun is shining bright;   *15) Background effects as available*

The band is playing somewhere, and somewhere hearts are light.

And somewhere men are laughing, and somewhere children

    shout;

But there is no joy in Mudville — mighty Casey has struck out.

From *Baseballmath: Grandslam Activities and Projects for Grades 4-8* published by GoodYearBooks. Copyright © 1994 Christopher Jennison.

"Casey at the Bat"   96

Leagues of their own

This project is a simplified version of the various forms of "Fantasy League" baseball that have become very popular in the past several years. It is recommended for only the most knowledgeable of your baseball fanatics. Basically a student "owner" selects twenty-two players from major-league rosters for his or her team, and after a predetermined period, the teams are ranked according to three batting and three

Skills in resource allocation, cooperative learning, problem solving, and calculation are used throughout this project. The only materials needed are copies of your Sunday newspaper's sports section in which the statistics of all major-league players are updated. If your local paper does not include this information, you can get it from the Tuesday edition of *USA Today* which includes American League statistics, and the Wednesday edition of USA Today for National League statistics. *Baseball Weekly*, a *USA Today* publication that comes out every Wednesday, includes totals for both leagues. Team "owners" can keep track of their teams' performances using the form provided on pages 99–100.

It is more realistic if student owners select their teams from the same division or league. But in order to make sure that there are enough players to go round, the number of student teams should be one less than the actual number of teams in a major league division or league. For instance, no more than four student owners should select from the American League East division team rosters, and no more than thirteen owners should select from all the American League rosters. The same would be true, of course, for the National League divisions and league.

Although the project can start at any time during the baseball season, and last

as long as you and the students decide, it is suggested that you begin on opening day, in early April, and finish in early June, after an eight-week season. Beginning in September would only allow a one-month season.

Let's assume that you have selected four student-owners, and that they will select their twenty-two-man rosters from the American League East division. The first thing that occurs is the draft, and this should take place just a few days before the major league season begins. Owners will select their players from opening-day rosters that may appear in your Sunday newspaper just before opening day. If not, the official roster will appear in USA Today or Baseball Weekly issues that are published just prior to opening day. Each owner has a salary budget of $22.00. (Real money is not used.) Minimum salaries are ten cents, and the minimum bidding increment is ten cents. Make sure owners keep track of what they spend, and that they don't exceed their budgets.

A team consists of five outfielders, two catchers, one first baseman, one second baseman, one shortstop, one third baseman, one middle infielder (MI) (shortstop or second baseman), one cornerman (CM) (first or third base), one utility man (UM) (National League), one designated hitter (DH) (American League), and eight pitchers. Players must be eligible to play the positions for which they are selected. Eligibility is based on their appearing at that position in at least twenty games during the previous season. That information will appear in newspapers just prior to the start of the major-league season. Once the students' season has begun, player eligibility is based on the position played on opening day.

An owner begins the draft with a minimum bid for a player of ten cents. (It doesn't matter who starts the bidding.) The bidding then continues at ten-cent intervals until there is one bidder left. Owners need to be warned, of course, not to spend too much money on one player, or risk not having enough salary to fill out their rosters. Players not drafted go into the reserve pool, which is used later in the season to replace injured or traded players. The draft will take at least two to three hours, longer if more teams are involved, so it's best to schedule it for after school or the weekend before the season begins.

Once the major league season starts, the student-owners keep track of their players' performances using the form on pages 99–100. Box scores appear in daily newspapers and *Baseball Weekly*. Many Sunday sports sections provide cumulative statistics, as do *Baseball Weekly* and *USA Today*.

At the end of the first week, and once a week thereafter, the owners will need to hold an Exchange, during which an owner may release players (a) placed on a Major- League disabled list, (b) sent down to the minors, (c) traded to the other division or league, or (d) released by their "real" team. These players are replaced by players from the reserve pool who are eligible for the position they are replacing.

Injured players are placed in the reserve pool, as are players sent to the minors once they return to their major league teams. These players can be reclaimed during the next Exchange.

Transactions during Exchange take place in reverse order of the most recent team standings. Therefore, the last place team in the standings has the first option to replace a lost player. An owner can make only one transaction at a time, so if he or she needs to replace more than one player, the second transaction cannot take place until every other owner has had a chance to make a first transaction. Remind owners that players released into the reserve pool can be claimed by another team a week later. So, if a star player is injured, but predicted to be out of action for only a week or so, the owner might want to hold on to him, rather than give him up for a lesser player. The star could be reclaimed, but the previous owner might not have first option at that point.

There are three batting and three pitching statistics used to decide the winning team. Owners must keep track of these statistics daily so weekly team standings will be known on Exchange day. In a six-team league, the team with the most home runs gets six points, the second place teams gets five points, and so on. The same scoring method is used for the pitching records. In the unlikely event of a tie, the winner could be the owner with the most money left in his or her "account."

HINT: After reviewing the rules with your student-owners, you may want to name a "Commissioner" who would be responsible for enforcing the rules throughout the season.

From *Baseballmath: Grandslam Activities and Projects for Grades 4-8* published by GoodYearBooks. Copyright © 1994 Christopher Jennison.

Team name: _____ **Owner's name:** _____

Pos.	Player		Week 1	Week 2	Week 3	Week 4	Week 5	Week 6	Week 7	Week 8
OF		Avg.								
		Hits								
		HRs								
OF		Avg.								
		Hits								
		HRs								
OF		Avg.								
		Hits								
		HRs								
OF		Avg.								
		Hits								
		HRs								
OF		Avg.								
		Hits								
		HRs								
C		Avg.								
		Hits								
		HRs								
C		Avg.								
		Hits								
		HRs								
1B		Avg.								
		Hits								
		HRs								
2B		Avg.								
		Hits								
		HRs								
3B		Avg.								
		Hits								
		HRs								
MI		Avg.								
		Hits								
		HRs								
SS		Avg.								
		Hits								
		HRs								
CM		Avg.								
		Hits								
		HRs								
DH		Avg.								
(or)		Hits								
UM		HRs								
TOTALS										

TEAM NAME: _____ OWNER'S NAME:_____

Pos.	Player		Week 1	Week 2	Week 3	Week 4	Week 5	Week 6	Week 7	Week 8
P		W								
		SVs								
		ERA								
P		W								
		SVs								
		ERA								
P		W								
		SVs								
		ERA								
P		W								
		SVs								
		ERA								
P		W								
		SVs								
		ERA								
P		W								
		SVs								
		ERA								
P		W								
		SVs								
		ERA								
P		W								
		SVs								
		ERA								
TOTALS										

From *Baseballmath: Grandslam Activities and Projects for Grades 4-8* published by GoodYearBooks. Copyright © 1994 Christopher Jennison.

Answer key

Answers vary in activities for which answers are not provided below.

Place hitters
1. $8.00
2. 3
3. 70
4. $3.50
5. 9
6. 5
7. 3
8. 20

Card profits
1. made $.20
2. made $.15
3. made $.35
4. $1.00

Circling the bases
1. 360 feet
2. 240 feet
3. approximately 21.8 seconds

Change champ
1. $.75
2. $1.50
3. $.50
4. $.25
5. $1.75
6. $2.50
7. $.40
8. $2.90
9. $2.75
10. $2.00

Team trip
1. 1,073
2. 976
3. 582
4. 875

5. 1,180
6. 1,457
7. 878
8. Chicago and St. Louis
9. Boston and Houston
10. Boston—it travels the most miles
11. St. Louis

Diamond data
1. 31 inches
2. 30 inches
3. Answers will vary.
4. 5'8"
5. 61 pounds
6. Answers will vary.

Sporting goods
1. $4.89
2. $.73
3. $12.49
4. $3.09
5. $.26
6. $2.95
7. $5.18
8. $1.36

A trip to the park
1. 30 minutes
2. 30 minutes
3. $2.40
4. $13.20
5. Answers will vary.
6. Answers will vary.
7, 8, and 9. Opinions should be supported.

A game of inches
1. Answers will vary in all cases.

Choosing sides
1. Answers will vary, depending on selections.

Graphing favorites
1. 5
2. 2
3. 8
4. Blue Jays
5. Rangers
6. Cardinals

Average attendance
1. 8,093
2. 8,729
3. 6,298
4. 7,127
5. 8,609
6. 9,388
7. 7,044
8. 6,403
9. More, since 10,926 is higher than average for first three games

Crystal (base)ball
1. No-hitter pitched by a chimpanzee
2. Player steals eight bases in one inning
3. Doubleheader is played on the moon.
4. Players agree to play for no money.

Big bucks
1. 25 billion
2. 16.8 billion
3. decrease
4. 8.2 billion
5. opinion
6. 4.2 billion
7. 4.2 billion
8. 4.2 billion
9. 2.5 billion
10. 1.7 billion

Hit parade
1. July 2, July 2, July 5
2. July 2
3. July 6
4. July 10
5. 7
6. Cardinals
7. Tigers
8. Indians

Time out
Answers will vary.

Baseball budgets
1. 7%
2. six out of every
    hundred
3. transportation
4. taxes
5. administration
6. insurance
7. player salaries
8. one out of every five,
    or twenty out
    of every hundred

Growth stocks
Estimates will vary.

Inning time
Estimates will vary.

Pitching pros
1. twenty
2. two
3. Glavine
4. 14
5. fifteen, twenty
6. eighteen

Where they stand
1. 6
2. 7
3. 7
The Cosmos will finish first,
    one game ahead
    of the Black Holes.

Slugfest
First   Hank
Second   Aaron
Third   Champ
Fourth   Homer
Fifth   Slugger

Game goodies
Answers will depend on
    estimates and
    will vary.

Lining up
1. 4
2. 12
3. 10
4. fifth
5. second
6. 44

Tips for tips
1. $2.00, $14.25
2. $9.00, $72.38
3. $1.00, $8.80
4. $45.53
5. $22.95
Note: Students may elect
    to round off totals too,
    which would be
    acceptable

Homer daze
1. 2
2. Wednesday
3. Tuesday
4. Wednesday
5. 50
6. Answers will vary, but
    anything between
    two and six would
    be reasonable

Collecting cards
1. Answers will vary.
2. 12, 24, 36, 48, 60, 72
3. 144
4. Approximately 70, depending on
    how student rounded off
    numbers

Wait 'til next year
1. Baltimore Orioles
2. Orioles' percentage in 1991 was
    .414; in 1992 it was .549,
    an improvement of 33%
3. Seattle Mariners
4. .512
5. 30%

Home run showdown
1. 16%
2. 33
3. 15 at bats
4. 143
5. a little less than 1

Better than the boys?
1. 27
2. about 36%
3. about 32,000 at 37%

The ballpark factor
1. Red Sox
2. Yankees
3. Dodgers
4. Braves

Playing favorites
Answers will vary.

 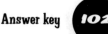

Bigger bucks
1. $4.6 million
2. $35,385
3. Answers to questions 3-6 will vary.
7. $2,500,000
8. at least $250,000 per season

Skill sharpeners
1. 8 hours, 40 minutes
2. 1 hour, 30 minutes
3. 45 minutes
4. 6 hours, 40 minutes
5. 2 hours, 50 minutes
6. 14 hours, 30 minutes

Uniform numbers
Kelsey: 7
Sonia: 11
Gary: 17
Lionel: 4
Chris: 100
Ellen: 25

Long arms
Carrie: 26.4m
Roy: 15.7m
Kyle: 21.18m
Ingrid: 26.34m
Eduardo: 26.34m
1. Roy
2. Ingrid
3. 13.33m
4. 8.23m
5. 36.46m
6. Ingrid

Picture players
1. 4 inches
2. about 9 2/3 inches
Answers will vary.

Striking out
1. Phil 1.73 to 1
2. Karen 1.13 to 1
3. .352
4. about 47

5. .294
6. about 59
7. Patty .22
8. Answers will vary.

Full house
1. 29
2. 60
3. 306
4. 105
5. 30
6. 20
7. 109
8. 70

Ballpark field trip
1. How many buses will be needed?
2. How many students will not be able to go to the game if only eleven buses show up?
3. How many seats are in the 23rd row?
4. The ballpark has 4400 seats. Each row holds 200 seats. How many rows are there?
5. The ballpark has 4600 seats. Each row holds 200 seats. How many rows are there?
6. Note: Other answers are possible.

Time for baseball
Answers will vary.

Finding the average
1. 6,8,5,7
2. Ronnie: 3, Jenny: 2, Tom: 3, Bobbie: 3, Julio: 4
3. 8530
4. Len: 9, Marian: 8, Sammy: 9, Chuck: 8, Stan: 5, Holly: 7, Michael: 6

Red hots
Hot dog: $.80
Scorecard: about $.45
Hats: $3.00
Ball: $4.90
Soda pop: 150%
Ice cream: $1.00

Low points
Butter Fingers: second
O. U. Stink: .066, fifth;
Shakee Batts: 63, .080, fourth;
Lowzee Hidder: 290, 6,889, .042, seventh
U. R. Badd: 4,117, 238, .058, sixth
Billy Goat: 328, 3,998, .082, third;
Hesa Whiffer: 6,632, 101, .152, first;
Bench Warmer: 22, 766, .029, eighth;
Lefty Failure: second;
Knuckles Flimsy: 77, 789, .098, third
Outa Control: 32/411, .078, fourth;
Bull Pen: 66/972, .068, fifth;
I. M. Crummy: 22,619, .036, sixth
Uneeda Curve: 71, 452, .157, first

Counting the crowd
1. American League
2. Atlanta
3. N.Y. Mets
4. up

Player targets
Ripken stands to break Gehrig's record in June 1995.